T0211666

The Business of Electronics

The Business of Electronics
A Concise History

Anand Kumar Sethi

THE BUSINESS OF ELECTRONICS
Copyright © Anand Kumar Sethi, 2013.

Softcover reprint of the hardcover 1st edition 2013 978-1-137-33042-0

All rights reserved.

First published in 2013 by
PALGRAVE MACMILLAN®
in the United States—a division of St. Martin's Press LLC,
175 Fifth Avenue, New York, NY 10010.

Where this book is distributed in the UK, Europe and the rest of the world,
this is by Palgrave Macmillan, a division of Macmillan Publishers Limited,
registered in England, company number 785998, of Houndmills,
Basingstoke, Hampshire RG21 6XS.

Palgrave Macmillan is the global academic imprint of the above companies
and has companies and representatives throughout the world.

Palgrave® and Macmillan® are registered trademarks in the United States,
the United Kingdom, Europe and other countries.

ISBN 978-1-349-46100-4   ISBN 978-1-137-32338-5 (eBook)
DOI 10.1057/9781137323385

Library of Congress Cataloging-in-Publication Data

Sethi, Anand Kumar.
    The business of electronics : a concise history / Anand Kumar Sethi.
        pages cm

        1. Electronics industry—History. I. Title.

HD9696.A2S48 2013
338.4'7621381—dc23   2013024804

A catalogue record of the book is available from the British Library.

Design by Newgen Knowledge Works (P) Ltd., Chennai, India.

First edition: November 2013

10 9 8 7 6 5 4 3 2 1

This book is dedicated to all those who so lovingly and tenderly looked after me during my recent face off with near death and extensive hospitalization. My family, in particular Deepa, whose abounding love, devotion, and care saw me through those very difficult times. The nurse, Sister Rajwant, who looked after me so well for days, in the intensive care ward always ready with not only injections, intravenous drips, electrodes, wires, tubes, and other instruments of torture, but also comforting words of great encouragement. Dr. Santosh Kutty, whose first timely intervention and assistance made it possible for me to fight for life. But above all, this book is dedicated to my doctor, Atul Joshi, without whose medical skills and ministrations I would not have lived to write this book. Undoubtedly, a superb surgeon but an even better human being!

Contents

Preface

Some years ago, an international television network flashed a fantastic photograph of an Indian ascetic merrily chatting on his mobile telephone while attending a Hindu religious event, one of the largest congregations of humankind held every six years. At that time, unsurprisingly, the photograph created quite a sensation. Today, given the all-pervasive presence of mobile phones, of all sizes and specifications used by global multitudes, such a photograph one suspects, would hardly raise any eyebrows.

Electronic gadgets and other appliances and systems using electronics ranging from the transistor radio, television, remote controllers, cameras, and vehicles through to the latest iPod, smart phones, and tablets have indeed become so ubiquitous that most of us do not give even a passing thought to what are the origins and the history of this life altering phenomenon called "electronics."

My own fascination with electronics began as a young boy growing up in a small Indian town in the early 1950s. Our first telephone was one where you had to crank a lever to gain attention of an operator at the telephone exchange. The diaphragms in the earpiece as well as the "talk" unit were the size of a carrom board "striker." Our music at home came from a hand wound "gramophone" turntable on which a thick Bakelite-based 78-rpm record was placed and a pointed needle stylus moved through the recording grooves giving out the music, courtesy of an apology of a "speaker" system.

But one day my world changed! Our old two-band mains-operated radio with a fascinating green colored tuning "eye" was replaced by a huge, unbranded, multiband radio set acquired as a surplus piece off a decommissioned World War II (WWII) vintage US naval ship. The world's radio stations broadcasting on shortwave frequencies were now my constant companions. Even more fascinating however was the sight of the multiple brightly glowing vacuum tubes, gang condensers, wire wound resistors, the ferrite rods, and the tuning coils that could be seen just by unscrewing the rear metal plate of the radio. What fun it was tweaking the antenna coils just to get that little bit extra tuning despite all the electrical shocks one received!

Some years later, I was successful at the entrance tests to obtain a coveted seat at the prestigious Indian Institute of Technology, Bombay, to study for a graduate degree in electronics. As a reward, my grandfather presented me with a Model 888 "Eight" Transistor (Germanium point contacts) radio set manufactured by the Emerson Radio and Phonograph Corporation. This radio, which I have kept repaired and operational for over 50 years, was my introduction to solid-state electronics. A love at first "sight" that endures until today!

The habit acquired of opening up and tinkering with electronic gadgets has endured. So much so that it almost got me into serious trouble on an official trip, as part of an electronics delegation to Pyongyang (North Korea) in the 1980s. The hotel "radiogram," a rather large contraption with a ten-band radio, tuned on to just one local station. I just absolutely "had" to open the set. Sure enough, in place of a gang condenser tuning unit there was a fixed-frequency crystal oscillator circuit inside. Worse, the room "bugging" system was placed inside the radiogram and an alarm was triggered as soon as the set was opened.

The growth of electronics ever since Thompson discovered the "electron" in 1897 has been, to put it mildly, quite spectacular! Helped in no small part by the growth of the "chip" technology following the trajectory of "Moore's Law," today we see some aspects of electronics in almost everything around us. Entertainment, telecommunications, and computing devices in their myriad versions and with equally diverse applications can today be found in the remotest corners of the world. Today, young children at the age of three or four work on tablet computers with a facility, that in our time we perhaps did not have even with paper and pencil. Yet, very few pause to ponder about the origins and the history of the development of electronics from a scientific curiosity into an all-pervasive, ubiquitous technology.

In the modern world and for the public at large, the term "electronics" tends to cover a wide spectrum of technologies ranging from conventional electronics and related products, telephones, audio-video systems, instrumentation, defense electronics, watches, and mobile phones through to magnetism, and the gamut of IT hardware and software. Each of these has a fascinating history. Thus, with due respect to purists of my tribe, I have taken the liberty in the following chapters to cover some aspects beyond the realm of classical electronics.

WWII was a time when the ingenuity and technological skills of the scientific community was severely tested. Electronics and telecommunications engineers too had to deliver, in many cases with technological development just to save one's own country and those of their allies, all to be delivered in a very short time. As is well known, some of the most significant electronic developments took place at that time. The computer and radar (Radio Direction and Ranging) are only two examples. Who may actually get the credit for being the first to make a certain product or come out with a particular technology is at times difficult to pin down with certainty. An effort has been made in this book to put forward the facts as are now reasonably well known. It is left to the readers to come to their own conclusions.

It is ultimately the many great scientists, engineers, inventors, "tinkerers," innovators, and entrepreneurs as also the hundreds of companies from many different countries that have made the most significant contribution to the development of what we now call electronics. Some names, justifiably, are now household names and icons in their own rights. Several others, sadly, are now forgotten and are part of history but do deserve an honorable mention.

Many of the pioneering companies and indeed research laboratories are still around and deserve all the praise they get. Yet others have floundered or simply disappeared from public memory. Some companies, great at one time, indeed technological and market leaders of their time, became "history" because of not foreseeing the future well enough. In some cases, the management of some entities took that one wrong decision at the wrong time with catastrophic results. This book does not do the analysis—history itself is the judge!

Developments over the years in electronics in all its manifestations, have taken place in many countries around the world and have involved contributions from countless companies, laboratories, research organizations, universities, and, of course, all the numerous individual contributors. Clearly, it would be impossible to mention and/or carry a history, however brief, of each and every one of them. In several cases there is a lack of proper documentation and hard evidence of what precisely transpired and their ultimate fate. Yet, care has been taken in this book to cover and mention as much as is realistically possible. Furthermore, many significant developments have taken place over a period of time and have involved contributions by many individuals, institutes, laboratories, and companies. With this in mind as also to make the book more interesting to the general reader, one has stayed away from a cut and dried "chronological listings" format in the chapters to follow.

Technological developments, inventions, research breakthroughs, and announcements of the introduction of latest electronic products and gizmos do fascinate one and all, including the purists. Many of these products, including some of the most popular consumer items, follow predictions made years earlier by science fiction writers such as Ray Bradbury (mobile telephony, large screen televisions, and iPods), and Arthur C. Clarke (iPad; refer to Chap. 9 of the book *2001: A Space Odyssey*)! However, the real and ultimate test for all of these items is that of the "market." Over the years, on the one hand, we have had products and technologies that looked exciting and promising but failed to "deliver" in the market place. On the other hand, we have also had products and technologies that, at first view, were not that "hot" yet went on to be the darlings of the market.

It is, however, the myriad companies engaged in the field of electronics that ultimately conceptualize, develop, promote, and take these products and technologies to the market at large. Many succeed and yet many others, including those established by the inventor(s) or the developer(s) of the products and technologies themselves falter, fail, and go into historical oblivion. Many did not quite make it possibly as a result of a failure to develop a good ecosystem around the product or the technology. It is the reading about some of these companies and businesses, much like that of towering personalities, statesmen, kings, and queens in conventional history that fascinates one and all. This book thus strives to highlight some of the fascinating stories about these companies, some successful, some failures, and some "also rans."

In conclusion, we hear reports emanating from the University of Southern California about the development of a microchip implant to help stroke victims. That is the good news! But we also hear talk of a day not too far into the future when a "wonder chip" with preloaded "apps" would be implanted into newborn babies so that in the future they would not be burdened with multiple electronic devices (a recent article in *The Economist* stated that RFID [radio-frequency identification] and other chips, externally positioned on children's wrists, bags, etc. are already very prevalent).[1] I sincerely hope that I am not around to record that development in the history of electronics, but then people like me do run the danger of being classified as foolish "romantics" from another day and age!

CHAPTER 1

Introduction

(Electronics) Engineering is merely the slow younger brother of Physics.
—Dr. Sheldon Cooper in the serial *Big Bang Theory*
(Season2, Episode 12)

It is the use in history of the Greek word "elektron" (formed by the sun), which gives us the etymology of the term. According to Rev. C. W. King,[1] elektron was connected to Helios, the Sun God, one of whose titles was "Elector" or the "Awakener." Collings[2] relates the mythological story that when Helios's son Phaeton was killed, his grieving sisters became poplars and their tears became the origin of elektron. According to Heilbron,[3] it is the work of William Gilbert,[4] of whom more later in this chapter, that showed that Amber could attract other substances and hence points one to the terms "electron" and "electricity."

But it was Dr. George Johnstone Stoney, an Irish scientist then at the Queens College, Galway, who in a paper in the *Transactions of the Royal Dublin Society* in the year 1891 first formally coined the term "electron." This was to become the forerunner of the many great developments in modern electronics.

So when did the whole "ecosystem" of "electronics" really begin? Almost certainly, the oldest known forms of communications over distances, by fires, drums, beacons, and smoke signals, existed in prehistoric days in possibly sixth century BC. We certainly know from Homer ("The Iliad") about fire signals in the Mycenaean period, which would be 1600–1200 BC. Communications yes, but not strictly electronic!

Magnetism, in the form of lodestone (magnetized mineral magnetite) was also known in sixth century BC with initial references to their properties made by Greek philosopher Thales of Miletus.[5] Most likely, the word "magnet" derives from "magnetis lithos," the lodestones found in Magnesia, Northern Greece. There is a popular legend relating to the discovery of the powers of lodestone. According to this legend, in 4000 BC, an elderly Cretan shepherd by the name of Magnes was herding his sheep in Magnesia when the nails in his shoes and the metallic top of his staff got stuck to the large black rock on which he stood. This rock contained lodestone.[6]

According to a fascinating article written by Ricker[7] and carried in the *General Science Journal*, the principle of the magnetic compass may have been in active use in China around 2630 BC, and further evidence is cited about compass usage in China in1100 BC.[8] This was of course to lead to the development of the Chinese mariner's compass, a sliver of lodestone floating on water during the rule of the Song Dynasty around the year AD 1100. The Olmecs of Central America may also have discovered the geomagnetic lodestone compass as a directional device as early as 1000 BC.[9]

Proponents of ancient Indian scientific achievements are proud to claim that as early as fourth century BC India not only had mastery of aeronautics and aircraft technology[10] but also had advanced knowledge of electronics and avionics. The above-cited work as also an ostensibly scholarly work by Dr. V. Raghavan,[11] a professor of Sanskrit language and possibly not a scientist, have references to, among others, the following:

Digpradarshana: An appliance to project the direction of approach of an enemy plane (radar?).

Parashabdha Graahaka Rahasya: Basic transmitter—receiver designs and frequency interception techniques of communications from enemy planes.

Vishwa Kriyaa Darpana: An electronic device used to obtain high-resolution real-time imagery of objects around an aircraft.

Shaktyakarshana Darpana: A device used as a protection against harmful nuclear radiation from weapons of mass destruction.

Vyroopa Darpana: A tactical device to project holographic image on a screen.

Adhrishya and Goodha: Devices to convert an aircraft into an invisible "stealth" machine.

Indian mythology particularly in the form of the holy book Ramayana describes aerial battles using such aircraft and devices. There is, however, no documentary and verifiable evidence in modern times that can confirm the existence in that day and age of such technologies. If indeed such technologies existed then the only logical assumption one can make is that there must have been a catastrophic destruction of all traces of these technologies, as certainly modern India, with the exception of the well-established and proven ancient India's contributions of the decimal system, algebra, trigonometry, and the first structured logical statements (now termed "software"), has had to develop or acquire electronic technologies much like the rest of the world.

Yet, the modern world does have clear evidence of the existence of technology for "batteries" dating possibly as far back as 2500 BC. Dr. Wilhelm Koenig, a German archaeologist and in 1938 the director of the National Museum of Iraq, proposed in a paper that some strange pot like artifacts with copper cylinders encasing iron rods in the museum collected from the village of Khuyut Rabbou'a near Baghdad were in fact galvanic battery cells and termed the "Baghdad Battery" or "Parthian Battery."[12] The Parthians were known to be in occupation of that part of Iraq from about 250 BC to 225 BC. In the same museum, Dr. Koenig also found some copper vases excavated from Sumerian sites dated back to at least 2500 BC, plated with silver, and concluded that this could have only come from an electroplating process possibly using vinegar or grape juice as an acidic electrolyte. Several scientists around the world have subsequently conducted experiments to recreate the working of the Baghdad Battery with some reasonable success. However, no such artifacts were ever found with conducting wires thus making it difficult to confirm whether such batteries actually predated the actual modern-day battery invention in the year AD 1800.

The ancient Indian text called the Agastya Samshita written by the sage Agastya, dating back to the first millennium BC, describes the operation of a battery cell to separate the constituents of water into its constituent gases. There are similar stories from other parts of the world. The aluminum found in the girdle inside the tomb of a Chinese General Chu (200 or so BC) could only have come from a battery-based electrolytic process converting bauxite. We also have Plato alluding in "Timaeus" to battery-operated lighthouses at Faro (Pharos). Mention may also be made of the light in the Temple of Venus/Isis as recorded by St. Augustine.

Ancient Egyptian texts, also tell us that at least as far back as 2750 BC, the Egyptians had some concept of "electricity" largely through shocks received from electric fish. Ancient writers such as Pliny the elder also testified to electric shocks from catfish. Around 600 BC, Thales of Miletos noticed static electricity and unfortunately wrongly believed that rubbing (static charge) made Amber magnetic.[13]

Thales was, however, not to know that many years later there would actually emerge a close relationship between electricity and magnetism, with a practical demonstration (the deflection of a magnetic needle placed close to a current carrying wire) by Hans Christian Oersted of Denmark and a mathematical corelationship developed by Andre Marie Ampere, in 1820. Ampere then went on to demonstrate the precise nature of the relationship between electricity and magnetism and to formulate his now famous Ampere's Law of Electromagnetism.

Solar technology was also known possibly as early as seventh century BC when magnifying glasses were reportedly used for making fire. Ancient texts suggest that the Greeks and Romans used mirrors to light torches for religious and other purposes. Archimedes, we know, as early as 212 BC used reflecting bronze sheets to focus sunlight to set fire to marauding Roman ships at Syracuse. Chinese documents also indicate the use of mirrors to light torches. But clearly, this was just passive solar technology.

Yet the rationalists may say, rightly perhaps, that there is no real clinching evidence that sophisticated technologies really existed in ancient times. Or may we let the romantics continue with their beliefs? Perhaps it is best to have the late great Carl Sagan have the last word on this subject. He is famously quoted as having said, "The absence of evidence is not the evidence of absence."[14]

Let us, however, concede the Greek origin of the term "elektron" and move on to more modern times where we have clear, unambiguous, documentary, and clinching evidence of technological developments. It was in AD 1600 that the English scientist William Gilbert, who closely studied electricity and magnetism, showed the difference between static electricity produced from rubbing amber (elektron) and the magnetic effects from magnetite and lodestone. He is then believed to have coined the new term "electricus" (of amber or elektron)[15] leading to the present-day term "electricity."

In November 1745, a German scientist, Ewald J. von Kleist while experimenting with electricity, most likely produced by electrostatic induction or an early version of what we now call a Van de Graaf Generator, accidentally touched the generator to a nail stuck into the top of a medicine bottle through a cork and received a massive shock when he touched the nail. What he had found was the first device capable of storing charge (electrons), today described as a capacitor.[16] This discovery was subsequently independently confirmed by a Dutch scientist, Pieter von Musschenbroek in 1746. Since Musschenbroek was from the Dutch town of Leyden, such a device was called the "Leyden jar" and more practical versions of this were to be used extensively in later years by Benjamin Franklin.

Suffice it to say that consistently and safely producing electricity in those early pioneering years was far from easy. As a result, electricity required for any experimentation was not readily at hand. However, in 1800 a brilliant Italian scientist called Alessandro Volta, came up with a fantastic new device. He soaked cardboard in a good quantity of salt water (brine). He then placed alternating discs of zinc and copper electrodes separated by layers of the brine soaked cardboard as a stack. This "pile" later to be called the "voltic pile" was able to produce sufficient current especially by connecting several of these piles together. The world then had its first real *battery* although the term itself was first coined in 1748 by Benjamin Franklin to denote an array of charged glass plates. Some are inclined to give some credit for Volta's discovery to another Italian scientist, Luigi Galvani. It may, however, be pointed out that Galvani's work was more to do with human nerve impulses when subjected to electrostatic sparks. He famously demonstrated the twitching of frog muscles when a charge was applied from a Leyden jar.[17]

It was, however, the genius of Michael Faraday that was to give us the first practical capacitor capable of storing a charge and delivering it when required. Faraday was born in 1791 into a poor English family. He was self-taught as his family could not afford him a formal education, and started professional life as a chemical assistant at the Royal Institution. In a spectacular career in chemistry, he discovered among other items, benzene, the Bunsen burner, the modern laws of electrolysis, and the liquefying of gases. But his contributions to electricity and electronics are perhaps more spectacular. In a short period of time, he had discovered the fundamentals of electromagnetic induction, mutual inductance, the Faraday Cage, and in 1837, the first practical capacitor, and finally, the variable dielectric capacitor. No wonder then that the unit of capacitance, the "farad," has been named after him.

In this chapter, we have already noted the vital contributions of Oersted, Ampere, and Faraday in the field of electromagnetism. But in 1825, William Sturgeon another self-taught physicist, and lecturer at the East India Company's Military Seminary in Surrey, UK, was the first to demonstrate a practical *electromagnet* that in different forms would later become a vital component in electronic products.

In 1827, a German scientist and mathematician, George Simon Ohm, noticed that different materials had varying capacities to conduct electricity. In addition to external parameters such as temperature and humidity, this behavior of materials, he noted, had something to do with their inherent type and structure. Ohm named this property as "resistance." Ohm, while working with electrochemical cells of the type discovered by Volta, showed that there is a direct proportionality between the voltage applied across a material and the resultant current that flows, with the proportionality dependent on the resistance of the material used. This became known as the famous "Ohm's Law" although sometime later, in 1879, it was found that Henry Cavendish, the great British scientist (the discoverer of Hydrogen) had already discovered this phenomena but surprisingly had failed to publish his findings.[18]

Interestingly enough, Ohm and others made different types of resistors using different materials but it was a prolific African-American scientist, Otis Boykin, who is credited with a 1961 US patent for developing the modern resistor as we know it. Otis Boykin, though, is perhaps better known for his invention of the heart "pacemaker," a device that was to help many patients with cardiac problems.

Meanwhile, across the Atlantic Ocean another convert to science and technology but without a formal education was beginning to make some spectacular developments based on experimentation he would do working night shifts as a telegraph operator. Thomas Alva Edison, by the year 1877, even before his thirtieth birthday, had already invented a stock ticker and the phonograph.[19] Edison (with some help from one of his assistants, Nikola Tesla, who was to go on to become a great and famous inventor in his own right) then went on to give the world the "quadruplex" telegraph (transmission of four messages concurrently, two in each direction), the carbon telephone transmitter (microphone), the first incandescent bulb, the motion picture camera, the first modern electricity distribution system, and not forgetting the founding of the General Electric Company (GE, US), one of the great companies of all time (for more on Edison's companies, please see Appendix 2).

But it was Edison's work on glowing filaments and their properties that were to give electronics a big boost. By inserting a metal plate between two glowing filaments, he found that electricity would only flow from the positive side of the filament to the plate but not from the negative side. Not being a formally educated person, he could not explain this phenomenon. He had, however, unwittingly stumbled onto what was termed the "Edison Effect" and had made the world's first *diode*.

By the year 1865, the theoretical framework of electromagnetic waves had already been established. It was however left to Dr. Heinrich Hertz, a professor at University of Karlsruhe, Germany, to develop the Hertz radio wave transmitter using a high-voltage induction coil, a Leyden Jar, a spark gap as the so-called transmitter, and a simple "receiver" made of a copper wire bent as a circle with a brass sphere at one end.

This successful experiment was to lay the basis of an improved wireless telegraph, as well as the radio and the television in later days. The unit of frequency, hertz, was deservedly named in Hertz's honor.[20]

Prior to 1897, many scientists had worked on the model of the "atom." Conventional wisdom of the time held that although the atom comprised more fundamental units, such units were of the size of the smallest atom, those of Hydrogen. However in 1897, the great British scientist, Dr. Joseph John Thomson, at the age of 41, working on the properties of cathode rays found that not only were the cathode rays more than one thousand times smaller than the Hydrogen atom but also that their mass was the same irrespective of the type of atom they came from.

Thomson went on to conclude that the rays were formed by extremely light, negatively charged particles that were the fundamental building block of all atoms. He named these particles "corpuscles." It was only later that scientists applied the term "electron," first stipulated in 1891 by Dr. George Johnstone Stoney,[21] to these particles. The formal discovery of the electron was then to set in motion the modern development of what we may call the whole ecosystem of electronics.[22]

In 1904, another British scientist and interestingly another of Thomas Alva Edison's former colleagues, John Ambrose Fleming, then a professor at University College London, developed a double electrode (an "anode" and a "cathode," both terms popularized by Faraday) vacuum tube "rectifier," which he called the "oscillation valve." The cathode (negative electrode) when heated would emit electrons that would flow toward the anode (positive electrode or the "plate"). Electrons would, however, not flow from the anode toward the cathode as there were no thermionic emissions from an unheated anode. Later on, this device was to be renamed as the "thermionic valve" and also more appropriately, the "vaccuum diode" (the word diode originating from the Greek "di" for two and "hodos" for way).

An American physicist Dr. Lee de Forest, in 1906, added another electrode (the control grid) to the oscillation valve and created the vacuum tube radio frequency detector that he termed the "Audion" but became more popularly known as the "triode" (the word triode originating from the Greek "tri" for three and "hodos" for way). Lee de Forest found that by the addition of the third electrode, "the grid," he could amplify input signals of different frequencies. What he had developed then was the world's very first "electronic amplifier."

Lee de Forest established the De Forest Company, to produce his "Audions" but was unable to make it a success largely because his product was not a proper vacuum tube. He erroneously believed that the phenomena that he had seen somehow depended on there being some residual gas remaining in the tube. Yet, Dr. de Forest, a somewhat colorful person, forever in trouble over patent lawsuits, failed marriages, and failed companies is widely regarded as one of the founding fathers of modern electronics. With some three hundred patents, he is also sometimes referred to as "The Father of Radio."[23]

It was only in 1914, after an extraordinary Finnish inventor from Helsinki, Eric Tigerstedt[24] (also referred to as "Suomen Edison," or the Edison of Finland—more about him later in the book) had considerably improved the product, and made it practical and usable, while working on his favorite project of incorporating sound on movie films,[25] that the first commercially available triodes were subsequently produced by GE, Schenectady, under the supervision of Irving Langmuir,[26] who was later to receive a Nobel prize.

In 1914, Finland was technically a part of Russia, making Tigerstedt a Russian citizen, but since at the start of World War I (WWI), he happened to be based in Germany, he was thrown out of the country and his patents and other material confiscated and used by the Germans without Tigerstedt's permission and without any compensation.

With availability of knowledge about *electromagnetism* and the development of the vacuum-tube-based *diode* and *triode*, the *resistor*, and *capacitor* and *variable capacitor*, as well as the *battery*, the world now had the fundamental technologies and components required to assemble electronic equipment for telecommunications and telephony as also radio sets and other devices for use at home.

CHAPTER 2

The Early Years: Telegraphy and Telephony

The telephone is a good way to talk to people without offering them a drink.

—Fran Lebowitz

Ever since the Phoenicians and the Sumerians developed the basic alphabet, around 3000 BC, there has been an endeavor to improve communications between fellow human beings especially in terms of sending communications over some distances. We know that the Greeks used homing pigeons as early as 776 BC to send information to the Athenians regarding the winners of various events at the Olympic Games of that year. Much afterward, carrier pigeons were quite extensively and effectively used for carrying messages from battlefields in both the world wars.

Humans (especially runners) being used as messengers over distances is, of course, very well known. There is no better example of this than the 26.2 mile run in 490 BC by Philippides (sometimes also referred to as Pheidippides) from Marathon to Sparta to request help in defending Marathon against the invading Persian army of Xerxes. Of course, many hundreds of years later, runners were replaced by dispatch riders on horses such as the "pony express" before they themselves were replaced by messengers on wheeled vehicles. In several developing countries, "telegrams" are to this day delivered by persons on bicycles. Meanwhile, the venerable postal system, also now sometimes referred to as "snail mail," still carries on, although quite diminished in numbers of mail carried.

In 200 BC, it is believed that fire/smoke signals between repeater posts as communications were used in China and Egypt. Polybius, a Greek historian is believed to have developed a system of converting Greek alphabets into numerical characters and sending them by smoke signals. Popular folklore and movies have over the years informed us of Native Americans (Red Indians) using smoke signals for communications.

The use of drums was also a particularly good means of communications in the olden days. Communication drums were well known in Africa, particularly in West Africa and this concept spread to the Americas and the Caribbean during the era of slave trading.

In 1792, Claude Chappe from France with help from his four brothers developed a message signaling system for sending messages between Paris and Lille. The system used mechanical signaling arms on towers that were spaced some ten-kilometers apart.[1] The Chappe system was subsequently used across a large part of France and formed an integral part of its communications during the reign of Napoleon.

Flags, paddles, and disks were also extensively used in semaphore signaling and particularly so for communicating from ship to ship and from ships to shore and vice versa when in reasonably close proximity. Naval history books carry several references to the very successful use of flag semaphore communications at the Battle of Trafalgar. The use of lights for signaling, referred to as the "Aldis" lamp, was another technology used quite extensively by the British Royal Navy and is still prevalent in some navies especially when radio silence needs to be maintained.

In the more modern era, particularly so with the rapid growth of the railways, there was an imperative need to improve communications over longer distances not quite feasible using the flag-based semaphore system or the Aldis lamps. Early "telegraph" (derived from the Greek "tele" for "far" and "graph" for "writing") systems were made, possibly the first one in 1809 in Germany by Samuel Sommerling and in 1828 by Harrison Dyar in the United States who used a somewhat crude system of electrical sparks to burn dots and dashes into a chemically treated litmus-paper tape.

About the same time as Sommerling was experimenting with his version of a telegraph machine, a Russian baron (strictly speaking an Estonian from Talinn) by the name of Pavel L'vovich Schilling, with interests in Oriental literature and antiquities, was also doing his own experimentation on a telegraph. Sometime in 1830, he is believed to have held a demonstration of the equipment for the Russian Tsar in St. Petersburg and in 1836 a working system over five kilometers was installed for the Russian Admiralty. This system based on visual reception of codes, became a pattern for many of the electromagnetic telegraphs that followed. Incredibly enough, this system never received a patent, most likely because of Schilling's early death in 1837.[2]

The credit for the demonstration of a practical telegraph system is also sometimes given to Joseph Henry, an American scientist. In 1830, he managed to send an electric current over a one-mile length of wire to activate an electromagnet-based bell at the other end. Many years later, Henry was honored by the naming of the unit of "inductance" parameter after him.

It was, however, only in 1837 that Charles Wheatstone, a professor of experimental philosophy at King's College, London, and William Cooke formerly of the British Indian Army and subsequently a professor of anatomy, were able to patent in 1837 the first electric telegraph. This telegraph (also sometimes called the "universal telegraph") was subsequently to be improved upon and made cheaper by Samuel Morse, interestingly a professor of arts and design at New York University and not really a scientist, who publicly demonstrated a telegraphy unit by sending the now historically famous message "What hath God wrought" on it on May 24, 1844, from Washington, DC, to Baltimore.[3] Morse's message demonstration was to precede by a few months another famous message sent by telegraph, this time in Britain, informing *The Times* of London of the birth at Windsor Castle of Prince Albert, the son of Queen Victoria.

The Magnetic Telegraph Company (see Appendix 1 for a brief history of the company) promoted by Samuel Morse was incorporated in the United States in May 1845 and became arguably the world's very first telecommunications company. Morse became more interested in giving out licenses for his patented technology rather than manufacturing or providing services on his own. By 1851, there were over 50 telegraph companies operating in the United States, some of them using technologies developed by rival entities. One of Morse's licensees was the New York and Mississippi Valley Printing Company, which after reorganization in 1856 became the Western Union Telegraph Company (see Appendix 1 for a brief history of this company).

The Electric Telegraph Company registered in the United Kingdom in September 1845 is commonly and widely acknowledged to be the first public provider of telegraph services in the world after having purchased the patents of Wheatstone and Cooke. The Electric Telegraph Company merged with the International Telegraph Company also of Britain (which in 1852 had received landing rights for a communications cable between Britain and Holland) to become the Electric and International Telegraph Company[4] in 1855. This new company in 1860 went on to create a pioneering, long-distance international communications link between London and Calcutta, then the capital of British

India.[5] This company was nationalized in 1870 and was to be the forerunner of what we now know as British Telecom (see Appendix 1 for a brief history of the company), which has the distinction of being the world's oldest communications system provider.

Unfortunately, the development and commercialization of the electrical telegraph became beset with several controversies. First, Wheatstone and Cooke fell out with each other and parted company even before the Electric Telegraph Company had started commercial operations in 1846. Ultimately, the two erstwhile partners had to go in for arbitration. Records indicate that Samuel Morse while visiting Europe in the 1830s got to know about electric telegraphy as it was emerging as a promising technology in Britain and on the European continent. On returning to the United States, he conducted his own telegraphy experiments with some success. On his next visit to London in 1838, Morse applied for a British patent for his work but this was vehemently opposed by Wheatstone and Cooke and as a result, Morse was unable to get a British patent.

When Morse went on to France, so the story is told, he propagated his own version of the telegraph. It is, however, believed that the scientist put in charge of studying Morse's proposal, Jules Guyot, is said to have remarked, "What can one expect from a few wretched wires?"[6]

Finally, Morse made further improvements and developed what we now know as the Morse code. This differentiated his invention from those of others. In this device, an indelible record was made that differentiated it from the sort of "semaphore" version of the British duo of Wheatstone and Cooke; a US patent was received by Morse in 1847.[7]

Britain, of course, was not the only country in Europe where telegraphy had started to flourish. In 1847, Werner von Siemens of Germany set about improving the Wheatstone telegraph and could make it function over a distance of 50 kilometers. By the next year, in 1848 just three years after the establishment of Morse's Magnetic Telegraph Company, the company started by Siemens called Siemens & Halske (Johann Georg, the engineer partner) had established a five hundred kilometer telegraphy link between Berlin and Frankfurt, and by 1850, the company was already setting up a wide telegraphy network in Russia. Its crowning achievement of the time was the completion in 1867 of the Indo-European telegraph line providing another link between London with Calcutta in India (see Appendix 1 for a brief history of the company).[8]

In 1850, the British Electric Telegraph Company was formed to be a competitor to the monopoly of the Electric Telegraph Company. In 1853, this company merged with the European and American Printing Telegraph Company to form the British Telegraph Company. In 1852, a cotton merchant from Manchester, John Pender, had joined other businessmen to take over the management of the English and Irish Telegraph Company, running a telegraph service between London and Dublin. The British Telegraph Company and the English and Irish Telegraph Company merged in 1857 to form the British and Irish Magnetic Telegraph Company. This was to become the beginning of Pender's very successful involvement in the business of telecommunications and the forerunner of that other great British company, Cable & Wireless, which ultimately on July 27, 2012, became a wholly owned subsidiary of Vodafone.[9]

Messaging by telegraphy over wires and cables, including subsea cables as a form of international telecommunications thus became quite prevalent in the 1860s and also led to the formation of some of the great pioneering companies of the time, some still operational in somewhat different "avatars." The challenge now was to transmit and receive voice and audio signals over communication links. The world was now ready for "telephony" (from the Greek "tele" for "far" and "phony" for "sound") and the telephone, the instrument and technology that would enable telephony or communication of sound over distances.

An Italian inventor, Innocenzo Manzetti is generally credited with first mooting the idea of a voice-based telegraph as early as 1844. A French telegraph operator, Charles Bourseul, in 1854 wrote a paper in the *L'illustration Journal Universel*, explaining the principles of the electromagnetic and the printing telegraph, and the concept of a speaking telephone, although he does not seem to have actually made a working device.

It is, however, believed that the first genuine voice communication system (called the "Teletrofono") was built in 1857 by Antonio Meucci, an Italian immigrant settled in New York to enable him to communicate from his laboratory in the basement of his house with his invalid wife in the bedroom in the upper floor. Unfortunately, Meucci could not afford the filing fee for a patent for his device, so in 1871 he filed a notice or caveat for an "impending patent." Unfortunately, three years later he could not afford even the meager $10 required for the renewal of this notice. A fascinating article[10] appearing in the newspaper *The Guardian* claims that Meucci sent a working model and details to the Western Union Company who did not revert with any comments. Subsequently in 1874, Western Union said that the working model as also the documents were not to be found any more, and were presumed lost.

According to the above-cited article, "two years later *Bell (Alexander Graham)* who shared a laboratory with *Meucci*, filed a patent for the *TELEPHONE*, became a celebrity and made a hugely lucrative deal with

Western Union" (emphasis original). Meucci filed a suit against Bell and was believed to be near winning when he unfortunately passed away in 1889. It was only in the year 2002 that the US House of Representatives officially recognized the contribution of Meucci toward the development of the telephone, prompting the Italian newspaper *La Republica* to write "a belated comeuppance for Bell, a 'cunning Scotsman' and 'usurper' whose perfidy built a telecommunications empire."

The controversy regarding the actual inventor of the telephone is not just confined to the claims of Manzetti and Meucci. Elisha Gray, an inventor from Illinois, US, had established a company called Western Electric (see Appendix 1 for a brief history of the company), which was a supplier of systems to the telegraphy operator Western Union. Gray had developed a device in 1874 that could transmit musical notes over a conducting line to some distance, and dubbed the invention the "harmonic telegraph."

On February 14, 1876, Gray formally filed a caveat for a patent for his invention under the title of "Transmitting Vocal Sounds Telegraphically." Unfortunately for him, just a few hours earlier, Alexander Graham Bell's attorney had already filed a full patent application at the same Patent Office under the title of "Improvements in Telegraphy."[11] The patent was awarded to Bell, principally on the grounds that his application was received earlier and that it was a full patent application as against just a caveat filed by Gray. This was to lead to a protracted legal battle, which ended in Bell's favor. Interestingly, it was later to be found out that the item described in Gray's caveat would have worked while the same could not be said about Bell's apparatus.[12]

The invention of telephony also had a German contender. In 1860, Johann Philipp Reis, a schoolteacher with a keen interest in mathematics and physics, is believed to have constructed a working "telephon," as he called it. Unfortunately, this device had some shortcomings and never did function well. It then also failed to get recognition by a judge in a patent lawsuit.[13]

The great Thomas Alva Edison was also in the 1870s experimenting with products that could be linked to telephony, although in fairness his ideas were more in the line of a "loud speaker" than a working telephone. In 1877, Edison did go on to invent the "carbon transmitter," which was to become the most effective transmitter in telephony for many years. Further, that year Edison was able to demonstrate the sending of audio signals over wires using his newly developed transmitter and promptly filed for a patent, which was granted only in the year 1892.

Notwithstanding all the controversies and the legal imbroglio surrounding Alexander Graham Bell getting the patent for the telephone, his many contributions to telephony and electronics do deserve a very special place in history. Bell was born in Scotland in the year 1847. His family emigrated, first to Canada and finally in 1871 to Boston, US. His mother was hearing challenged (as later was one of his students and his wife to be) and his father was a tutor to children with hearing difficulties. It was then no wonder that Bell from an early age was very interested in helping the hearing impaired. He carried out extensive research and experimentation on the possibility of transmitting musical notes and audible speech over a distance.

In 1874, Bell chanced upon a meeting with Thomas Watson, an experienced electrical engineer and designer then working with the company of Charles Williams. Bell and Watson decided to work together for developing an acoustic telegraph, a project financed by Gardiner Hubbard, Bell's father-in-law. By 1875, Bell with Watson, working as his assistant, had developed an acoustic telegraph for transmitting sound using mercury as a transmitting medium (as against water used by Elisha Gray). On March 7, 1876, a patent was issued to Bell for his variable resistance telephone. It is believed that it was only after Bell had received the patent that he experimented with Elisha Gray's water-based design and it was the equipment using this design that Bell uttered across the line the now immortal words, "Mr. Watson, come here, I want to see you."

It was later alleged that the patent examiner in the dispute between Bell and Gray was an alcoholic and was in debt to Bell's lawyer, a former colleague of his from the US Civil War. It was also alleged that the patent examiner had passed on to Bell details of an earlier caveat filed by Gray containing details about his design and concept.[14]

Bell and his partners are believed to have offered the rights to the telephone patent to the Telegraph Company, the predecessor of Western Union, for the sum of $100,000 but were turned down on the grounds that the telephone was of no practical use. The purported comments of the evaluation committee alleged, possibly incorrectly, to have been carried in a 1968 edition of *IEEE Transactions on Systems, Man, and Cybernetics Society* read as follows: "Messers Bell and Hubbard want to install one of their 'telephone devices' in every city. The idea is idiotic on the face of it. Furthermore, why would any person want to use this ungainly and impractical device when he can send a messenger to the telegraph office and have a clearly written message sent to any large city in the United States."[15] The rest, as they say, is history!

The Bell Telephone Company (see Appendix 1 for a brief history of the company) was incorporated in 1877 by Gardiner Hubbard, Bell's father-in-law, and within a decade, more than 150,000 people in the

United States owned a telephone. Bell would then go on to collect several other patents for his inventions including those for the "photophone" (a device for transmitting sound over light), the phonograph, selenium cells, and the audiometer, as well as pioneering work on desalination and magnetic recording.

According to President Barack Obama, in 1876 when the then president of the United States, Rutherford Hayes, was first shown the telephone by Bell, he is believed to have remarked, "That is an amazing invention, but who would ever want to use one of them."[16] The veracity of the nineteenth president of the United States making such a quote is, however, doubtful! Yet, if President Hayes did make the alluded statement, how wrong he was! By 1900, there were some 600,000 telephones in the Bell system; by 1905, this number had grown to over 2 million; and by 1910, numbers were already touching 6 million.

While telephony was getting to be very popular in the United States, it was also catching on in Europe as well. We have already noted the advent of Siemens & Halske in Germany as a telegraphy operator. By 1877, Siemens & Halske were producing telephones that were an improvement on Bell's design by replacing the single bar magnet by a horseshoe shaped magnet. Unfortunately, Bell had not registered his patent in Germany enabling the German company to take full advantage of this lapse.

In 1876, a young Swede, Lars Magnus Ericsson, had established a telegraph repair workshop and then started to look at improving telephones made by Bell and Siemens & Halske. By 1879, Ericsson was already making his improved sound quality telephone, as well as for the first time combining the transmitter and the receiver into one composite handset and with this equipment came to dominate the Scandinavian market (see Appendix 1 for a brief history of this company).[17]

Telephone and telephony came to Britain as early as 1878 when The Telephone Company Ltd. was incorporated to sell telephones based on Bell's patents. In 1879, The Edison Telephone Company of London Ltd. was incorporated to put up telephony systems using Edison's technology. To begin with, the public in Britain were highly skeptical of this new-fangled technology, some preferring the good old-fashioned "air tubes" communicator or even ridiculously comparing telephony to British physicist Robert Hooke's 1667 version of what had been for many of us a fascinating childhood toy, the acoustic string communicator.

The telephone finally became popular in Britain after Bell demonstrated it to Queen Victoria and received the blessings of the famous scientist, Lord Kelvin (Sir William Thompson), for its great technological development. With imminent and expensive patent litigation looming and with a fractured market, the two British companies, perhaps sensibly, decide to merge to form the United Telephone Company Ltd. Another company, the Consolidated Telephone Construction and Maintenance Company was set up to undertake the actual manufacturing and servicing of the equipment. The two new entities were subsequently merged into what became the National Telephone Company whose assets were finally taken over by Ericsson in 1903 after a string of issues related to poor quality and run ins with government bureaucracy on the issue whether telephony was just another form of telegraphy, then under government monopoly.[18]

In 1880, the government of France bestowed on Bell the Volta Prize. The prize money that came along with this award was invested by Bell in establishing the Volta Laboratory, later to be called the Volta Bureau, where the pioneering works on the "photophone" and magnetic recording were carried out. Many erroneously assume that the Volta Bureau was the forerunner of what would in later days become the world famous Bell Labs. This great institution was set up by American Telephone and Telegraph Company (AT&T) only in 1925 and was named in honor of Alexander Graham Bell. Bell incorporated AT&T in 1885 as a subsidiary of the Bell Telephone Company, to provide telephony services. In 1899, AT&T fully acquired the Bell Telephone Company. It is perhaps also worth noting that Bell's major achievements with the technology of the telephone were made when he was still a British national. He became a US citizen only as late as 1882.

As has been noted earlier in this chapter, Edison had perfected a carbon transmitter, which enabled instruments using it to have a considerable edge over Bell's telephone. This was to cause no end of problems to Bell and more so to the then fledgling Bell Telephone Company. Fortunately for them, a German immigrant to the United States, Emile Berliner, a self-trained inventor, had developed a somewhat different form of transmitter working on a sort of loose contact between two metal electrodes, considerably increasing the volume of transmitted sound. Berliner had also filed a caveat for a patent for this device. The Bell Telephone Company was able to buy the rights to this invention and also hired Berliner, and thus saved the company from going under.[19]

But this was not to be the end of troubles for Bell. Early in 1887, the attorney general of the US government, Augustus Hill Garland, made an effort to get Bell's telephone patents annulled. This case had to be unceremoniously dropped after it was allegedly discovered that Garland had been given a very large number of shares in a company called the Pan-Electric Telephone Company, a competitor to the Bell Telephone Company and stood to gain immensely should Bell's patents be revoked.

Fortunately for science, electronics, and the world at large, Bell survived some five hundred court and patent challenges and went on to contribute many more scientific achievements to the world. At the time of his death in 1922, the entire telephone systems in the United States as well as in Canada are believed to have been shut down for two minutes in his honor![20] (For a listing of entities associated with Alexander Graham Bell and their timeline, see Appendix 3.)

In the early days of telephony, a connection between the caller and the called party would be made manually by cranking a handle on the telephone instrument, the signal from which would be received at a central station or "the exchange" as it became known later. Here, a bunch of "operators" usually ladies with nimble hands and fingers would ask the caller who he/she would like to speak with and then make a physical connection using a patch cord. The story goes that in 1899, an undertaker in Kansas City by the name of Almon Strowger noticed that a lot of his customers were getting diverted to his principal competitor. Strowger suspected that the competitor's wife, who worked at the "exchange," had a hand in diverting calls meant for him.

Strowger along with his associates then decided to develop a system that would do away with the manual patching in of telephone calls. By 1891, the new "switching" system called the "Strowger" system had been patented and the Strowger Automatic Telephone Exchange Company was formed to commercialize the patent. In 1901, the technology was licensed to the Automatic Electric Company, based in Northlake, Illinois. The company General Telephone and Electronics (GTE) acquired Automatic Electric Company in 1955, and ultimately became a part of Alcatel Lucent.

An even further development in telephony came with making the "switching" function fully electronic, increasing the speed and reliability of connecting telephone calls between subscribers. Several companies started to make such systems, chiefly Western Electric, Ericsson, Siemens, Northern Telecom, and Lucent. This was a normal progression of technology, first with the availability of what were termed "reed switches," two metallic terminals encased in a vacuumed small glass tube, and activated by an electromagnetic field. Subsequently, these were replaced by full solid-state electronic switching using integrated circuits (ICs; see chap. 9).

One of the leaders in making electronic public branch exchange (PBX) systems, was the US company Wescom Switching. A remarkable Indian engineer, Satyam (Sam) Pitroda joined this company in 1974 and led the efforts of this company in making a fully electronic stored program switch using, for that time the latest in solid-state devices, the microprocessor (see chap. 9), and the very new concept of "redundancy" in its circuitry. Unfortunately, Wescom Switching was soon to run out of funds and was sold off to Rockwell to carry the program forward. Sam Pitroda was later to become the father of the telecom revolution in India as well as to guide India's own efforts in developing an indigenous electronic telephone switching system. Subsequently, he was also tasked by the government of India in guiding the development of its nascent high-technology industries as well the country's innovation program.

Many years later, telephony would undergo some radical technological changes. It would be that it was not necessary any longer to be tethered at the other end of a telephone wire to make or receive phone calls. A subscriber would use a radio link for connectivity into a telephone network made up of several individual "cells." We will read about this concept of "cellular mobile telephony" in chapter 7 of this book. We will also read later, in chapter 8, about another radical development—telephony over a nearly globalized voice and data communications connectivity network, laid out more or less like an intricate spider's web pattern—the "World Wide Web" of the Internet.

CHAPTER 3

Wireless and Radio

The wireless telegraph is not difficult to understand. The ordinary telegraph is like a very long cat. You pull the tail in New York and it meows in Los Angeles. The wireless is the same, only without the cat.

—Albert Einstein

It's not true I had nothing on! I had the radio on!

—Marilyn Monroe

There was a time that the only thing you got from Japan was a really bad, cheap transistor radio that some Aunt gave you for Christmas.

—Cher

Telegraphy and Telephony brought about phenomenal changes in the way people were able to communicate with each other, send and receive messages. Yet clearly, there were shortcomings and drawbacks especially as communications between two points required a physical connection between them of an electrically conducting wire or cable. It was thus impossible to communicate telegraphy or telephony signals to remote places or even semiurban communities where wired connections were not available. In particular, it was extremely difficult for ships at sea, after the Industrial Revolution, ever increasing in numbers and traveling large distances, to communicate from ship to shore as well as ship-to-ship when they were not in visual range. The old semaphore and Aldis lamp systems were simply quite inadequate. A major technological breakthrough was now needed!

This came in the form of the wireless. Unfortunately, if one thought that the whole controversy about the real inventor of telegraphy was murky, the history to this technological breakthrough, one of the most exciting and significant technology developments of all times, was even murkier!

In the 1860s, the Scottish-born mathematician and physicist, James Clerk Maxwell, then already a scientist with great renown, having done pathbreaking work on photoelasticity, and the physics and color of light, joined King's College in London and started research on electromagnetic waves and electromagnetic induction. By 1865, Maxwell had enunciated a clear concept of electromagnetism and the theoretical basis of the propagation of electromagnetic waves.[1] This discovery of the nature of electromagnetic waves would then set the very basis for the development of radio, television, and radar, as well as, many years later, the mobile telephone.

In 1875, Thomas Alva Edison announced that during his experiments on telegraphy he had noted that if any metallic object touched the vibrating device he was using to generate high-frequency alternate current, it would produce sparks. Not able to explain this phenomenon, he named it as the "Etheric Force."[2] However, following criticism from fellow scientists, Edison dropped further work on this phenomenon, although we now know that what he had witnessed were actually radio frequency waves that generated sparks in metal objects in the proximity. Edison thus had come so very close to claiming the basis of radio communications as another one of his many achievements.

David Edward Hughes, born in Wales and went on to the United States when his parents migrated there, had already invented the printing telegraph system in 1855, which Western Union had put into widespread use in the United States. Further, Hughes improved on Edison's carbon telephone transmitter, which he called the "microphone." In 1878 while carrying out experiments, Hughes noticed that the sparks that were generated during experimentation could be clearly heard by an instrument fitted with his new microphone.

In 1884 an Italian inventor, Temistocle Calzecchi-Onesti, found that the resistance offered to electric current by iron filings inside a glass tube was considerably reduced in the presence of a radio frequency, which made the iron filings cling together letting current pass. This was what became known as the "coherer" effect and at that time formed the basis of devices to detect radio frequency signals, A French scientist, Edouard Branly, in 1890 went on to improve on the effect shown by Calzecchi-Onesti. He termed his invention the "radio-conductor."

It was, however, Oliver Lodge, a professor of physics at University College in Liverpool, England, who in 1894 formally coined the term "coherer." Lodge further went on to add a mechanism of "de coherence" to return the glass tube containing the iron filings back to its nonconducting state, which made the device much more sensitive. In 1898, Lodge received a patent for what was described as "syntonic tuning," which for the first time clearly spelt out the principle of "tuning" a radio circuit by varying the inductance of an antenna coil.[3] Unfortunately, Lodge did not carry on with his experiments with electromagnetic waves and thus missed out on becoming the inventor of "radio."[4]

Meanwhile in the early 1880s, Edison, who as we have noted earlier in this book was a telegraph operator in his early professional life, was working on trying to solve the problem of communications between two running trains on parallel tracks as well as between trains and the stations they passed along the way. He came up with a sort of "wireless" telegraphy, which was in fact based on the principle of electrical induction. A Morse code signal generated from a running train would have its voltage increased by an induction coil and the signal "transmitted" from a metal plate placed on top of a train could be picked up by telegraph poles with wires running parallel to the rail tracks.

Edison was to receive a patent for this "inductive train telegraphy" some four years after he had started work on the system.[5] Edison would later adapt this system for communications between ships and from ship to shore, and managed to sell this patent to Marconi (see below) but the system still remained essentially based on induction.

Between 1886 and 1888, Heinrich Rudolf Hertz worked actively on the theories of Maxwell and was successful with experiments validating the principles of electromagnetic waves, which he now termed "Hertzian waves."[6] Incredibly, even Hertz did not carry his experiments forward to a logical conclusion into the development of radio. In fact, on being asked what practical use his experiments would have, Hertz is reported to have said, "I do not think that the wireless waves I have discovered will have any practical application."[7]

From 1885 to 1893, at least two others laid some claim toward the development of the radio (Etymology—"radiate" and derived from "radiogram," the original term for wireless telegraphy). One was a farmer from Kentucky, Keith Stubblefield, who, as well as his descendants and people from his hometown of Murray, long proclaimed him to be the inventor of radio, but reports of the day seem to indicate that his transmission of signals was also based on the phenomenon of inductance.

There is then the story of a Brazilian Jesuit priest, Roberto Landell de Moura. He came to Rome to study theology but became fascinated with the reports of various experiments using electromagnetic waves carried out by leading scientists in Europe and America. In 1893, he is believed to have made a working model of a wireless telephone system and later demonstrated it to the bishop of Sao Paulo, Brazil. The bishop on hearing a sound emanating from the contraption is said to have accused Landell of practicing witchcraft and desired that further experiments be ceased forthwith. Some days later, Landell's laboratory was completely vandalized. Landell did manage to get a Brazilian patent for his invention but was unable to convince a patent examiner in the United States about the originality of his invention.[8]

A young Yugoslav, a graduate from the University of Prague, Nikola Tesla, used to work in Paris at the Continental Edison Company where he designed electrical dynamos. Not being able to get anyone interested in his unique design for an induction motor, he accepted an offer to come to work for Thomas Edison in New York in the year 1884.[9] Unfortunately, some years later they parted company on a major disagreement whether the future of electrical power and appliances lay with direct current as propagated by Edison or with alternating current as propagated by Tesla and supported by others such as the famous inventor and industrialist George Westinghouse.

By 1891, when Tesla invented the Tesla coil, now in widespread use in many electronic appliances, he had already been known for his development of the modern alternating current electrical motors, multiphase transformers, alternating current dynamos, and more. Tesla, arguably the greatest inventor in the field of electrical and electronics engineering of his time, went on to register some seven hundred global patents including those for fluorescent lights, lasers, bladeless steam turbines, and many developments in robotics, solar energy, energy from waves, and much more.[10]

Tesla began his work in wireless telegraphy in 1891. By 1893, he was able to publicly demonstrate in St. Louis, Missouri, a very basic wireless communication system. The system he devised had pretty much all the components needed to make a radio transmission prior to the actual use of what was to come a little later—the vacuum tube valve. By 1896, Tesla was able to construct an appliance to clearly receive radio transmission waves and had published details including the circuit diagrams, and filed for a patent in 1897. It is alleged that Guglielmo Marconi (see below) was later on able to use all of the details of Tesla's work for his own benefit.

Tesla's version of the transmitter generated radio frequency electromagnetic waves using a spark gap. This technology was then licensed to the Lowenstein Radio Company for use on board ships. The Lowenstein Radio Company would then go on to manufacture this product and several others largely for use on ships of the US Navy. Interestingly, Tesla, like Edison, never did receive a Nobel Prize for the great work he had done and his enormous contributions to technology and society. It may also be pertinent to mention here that as early as the year 1909,

Tesla in an interview to the *New York Times* had predicted the advent of mobile phones.[11] A great man clearly denied his due!

In 1885, the Indian scientist Jagadish Chandra Bose had returned to his hometown of Calcutta (now also called Kolkata) to teach and to do research, having received degrees from the University of Cambridge as well as from the University of London. Bose, who had initially gone to study medicine in the United Kingdom, switched to studying science and at Cambridge was taught by a galaxy of eminent teachers including the famous Lord Rayleigh. In 1895, Bose, in the presence of the then lieutenant governor of the state, demonstrated in the Calcutta Town Hall an experiment by which he could ring an electric bell placed at some distance by using electromagnetic waves.

Later that year, Bose was to improve on Lodge's coherer detector and in 1899 announced the development of an "iron-mercury-iron coherer with telephone detector" in a paper presented at the Royal Society, London. Some reports suggest that among the audience at the Royal Society lecture was Marconi himself; further, reports suggest that Bose's papers including his circuit diagrams were stolen from his London hotel room. Since at that time, Bose, being an idealist, believed that his scientific works should be for the benefit of all society, he was unwilling to patent his invention. Allegations then arose that Marconi took full advantage of the situation and was able to get a patent more than a year after Bose had made public his invention.

In 1901, Bose in a letter to his close friend, the great Indian poet and writer, Nobel laureate Rabindranath Tagore, wrote that shortly before his lecture at the Royal Society he was approached by a senior functionary of a very famous telegraph company and offered a very lucrative financial deal in exchange for jointly patenting Bose's inventions.[12] It is believed that Bose's interlocutor was none other than Stephen Flood Page, the managing director of Marconi Wireless and Telegraph Company.[13]

Bose subsequently went on to do pathbreaking research in many fields including plant physiology but in terms of electronics he developed a galena (natural mineral form of lead) detector for electromagnetic waves, and received a US patent for it in 1904. Hence, Bose may rightfully be called the father of solid-state electronics, if not that of the radio!

There is another legitimate contender for the title of "the father of the radio." Alexander Stepanovitch Popov, a professor of the University of St. Petersburg. Popov started work in the wireless field to develop an efficient detector for advance warnings of thunderstorms and lightning strikes. He selected Branly's radio conductor version of the coherer for his experiments. By 1894, Popov had built a working radio receiver using a modified coherer, and demonstrated this publicly on May 7, 1895, to the Russian Physical and Chemical Society. In 1896, Popov demonstrated at another meeting of the same society at St. Petersburg University how his device may be used for sending and receiving radio signals. In the demonstration, he transmitted the words "Heinrich Hertz" over a distance of 245 meters between two buildings of the university.[14]

By January 1890, Popov had set up a 47-kilometer radio link between Hogland Island in the Gulf of Finland and the present-day Finnish town of Kotka.[15] The Russians to this day insist that Popov is the real inventor of a practical radio system and May 7, the anniversary of the first demonstration of Popov's device, is called "Radio Day."

Notwithstanding all the outstanding contributions from the various scientists and inventors described above, in conventional terms the honor of being the inventor of radio is generally given to Guglielmo Marconi, the scientist from Bologna, Italy. Marconi was deeply impressed by the work of Heinrich Hertz and was confident that wireless radio was the means of signal communications over distances. By 1895, he had developed a system that could transmit a signal over a distance of over a kilometer. In 1896, Marconi went to Britain to apply for a patent for his system of wireless telegraphy and the patent for "Improvements in Transmitting of Electrical Impulses and Signals and in Apparatus Therefor" was awarded to him later the same year.

In 1897, then only 22 years of age, Marconi successfully demonstrated the capability of transmitting signals across a 14-kilometer stretch of the Bristol Channel, and in October the same year, he managed to transmit a signal over 50 kilometers from Bath to Salisbury, England. By November of 1897, Marconi had established the first-ever "radio transmission station" in a hotel located at Alum Bay on the Isle of Wight.

In the year 1899, Marconi was able to send signals across the English Channel from Britain to France and the following year he received another British patent for what was described as "tuned or syntonic telegraphy," which made it possible for multiple simultaneous transmissions using different frequencies.[16] Greatly encouraged by these successes, Marconi in the year 1898 had established his company, the Wireless Telegraph and Signal Company Ltd., which was subsequently, in 1900, renamed as Marconi Wireless Telegraph Company Ltd. (see Appendix 1 for a brief history of this company).

The major breakthrough for Marconi was, however, the purported transmission in 1901 of signals from Cornwall in Britain to St. John's in Newfoundland, Canada, a distance of about 3,500 kilometers proving clearly that his radio signals were unaffected by the curvature of the earth. This test, however, was to give rise to a controversy. It is believed that the only transmission made was the Morse code for the letter "s," represented by three identical "dot-dot-dot" and was carried out during daytime hours, which we now know is the worst part of the day for any transmission in what would have been medium wavelengths.

Marconi was, however, able to silence his critics when in the following year he could demonstrate the reception of the signals from Cornwall onboard a ship sailing on the Atlantic at a distance of well over 3,000 kilometers. On being informed of Marconi's achievement, the great Tesla is believed to have commented, "Marconi is a good fellow. Let him continue. He is using seventeen of my patents."[17]

A further controversy has been documented regarding Marconi. In the experiment relating to transmission of signals to St. John's Newfoundland, Marconi used a mercury-based coherer to receive the signals in conjunction with an earphone. Marconi called it the "Italian Navy Coherer." Now, this type of coherer is precisely what Bose had invented and demonstrated in London, and subsequently his papers and circuit diagrams were stolen, as noted above. When Marconi was questioned about this, he stated that he had received the design from an Italian Navy engineer called Solari, who, however, vehemently denied it. Marconi then said that the design of the specific coherer was obtained from an Italian, Professor Timasina, which statement was also later found to be false.[18] It is then very likely that Marconi knowingly used Bose's coherer design for his own experiment and did not wish to own up doing so!

Marconi was then to go on to build a radio telegraph station in a place called South Wellfleet in Massachusetts, US, which would then find a place in history not only in 1903 when the first transmission of a radio signal across the Atlantic from the US side was made (carrying greetings from President Theodore Roosevelt to King Edward VII) but also more importantly in 1912 when David Sarnoff, then an employee of the Marconi International Marine Company based at this station, is believed to have picked up the first distress signals from the mail ship *Titanic*. Sarnoff was reportedly the one to keep communications going assisting in the rescue operations by the ship *Carpathia*.

There is a further interesting story. It is believed that since Marconi's company was looking after the wireless telegraphy systems on the *Titanic*, he was offered a free passage on the liner on its fateful voyage, but for some reason, and mercifully for global electronics, he declined and sailed to New York on the *Lusitania* just three days earlier.

Marconi, although an inventor, was unquestionably also a very shrewd businessman and was quite willing to use his family connections with

British nobility to take his scientific endeavors and businesses forward. He established in 1898, arguably, the world's very first radio factory, the Marconi Radio Factory located in Chelmsford, Essex, in England, thus leading Chelmsford to being called "The Birthplace of Radio." In 1912, Marconi went on to build a purpose-designed new factory at New Street (since then renamed as Marconi Street), Chelmsford and from this location in 1920, a concert by Dame Nellie Melba was broadcast and heard in many parts of the world. Sadly as we write, the building has been lying unused since 2008 and quite derelict, including Marconi's old office room.[19] (For a listing of entities associated with Marconi and their timeline, please see Appendix 4 of this book.)

At the beginning of the 1900s, there was considerable wireless telegraphy related activity building up in Germany. There were two main groups that were doing advanced research; one was related to the Siemens & Halske group under the guidance of Dr. Karl Ferdinand Braun from the University of Strasbourg and another group was working on behalf of the German company AEG (Allgemeine Elektrizitatsgessellschaft).

At the behest of the then Kaiser, Wilhelm II, the two groups were merged in 1903 into the entity "Gesselschaft fur drahtlose Telegrafen mbh," Telefunken (see Appendix 1 for a brief history of this company) for short. By 1903, the Germans were successfully making wireless telegraphy transmissions of their own. Dr. Braun, in the meanwhile had received patents for his version of a crystal diode as also for tuning technology for wireless telegraphy. Several of these patents were also used by Marconi.[20]

Marconi in 1909 received the Nobel Prize, jointly with Karl Ferdinand Braun, for his contributions to the development of wireless telegraphy. However, in June 1943, the US Supreme Court invalidated Marconi's patent in the case of *Marconi Wireless Telegraph Company Ltd. versus the United States*, and most of his critics, and there are many, say he deserved this rebuke! The Supreme Court's decision clearly then went in favor of Tesla. Unfortunately, the Nobel Foundation was perhaps not in a position to callback the Nobel Prize already conferred on Marconi.

However as discerning readers would note, most of the so-called activities described above were still in the nature of wireless telegraphy transmitting Morse-code-type signals but not "voice." Stubblefield's experiments perhaps did have an element of voice audio transmissions, but there is reasonable certainty that this was based on an inductance principle and not on true radio signals.

Reginald Fessenden, a Canadian scientist who had earlier worked with Edison and Westinghouse, became head of the Electrical Engineering

Department at the University of Pennsylvania where he initiated further research on Hertz's experiments. The achievements of Bose, Popov, and Marconi spurred him on to develop a system that would carry actual audio voice signals and not just those in Morse code.

In the meantime, Fessenden had set up the National Electric Signaling Company (NESCO) in association with some millionaire backers and several of his patents were transferred to this company, which specialized in building transmission towers. But Fessenden was very keen to take an active role in wireless telegraphy in his home country, Canada. To pursue this objective he established in 1906 the Fessenden Wireless Telegraphy Company of Canada. It appears that Fessenden was not an astute businessman and his partners took full advantage of his naivety and exploited his patents for their own benefit.[21]

But Fessenden was not one to give up easily and was determined to carry his work forward to actually transmit audio signals. On the evening of December 24, 1906, Fessenden was able to broadcast from near Boston, Massachusetts, not only his voice announcing the subsequent program, but also the actual playing of Handel's "Largo" and himself playing "Oh Holy Night" on the violin. This broadcast, the first-ever known audio signals, was clearly picked up by several ships at sea. Meanwhile, Fessenden's problems with his partners in NESCO continued and resulted in his being ousted from the company.

NESCO was acquired by Westinghouse (see Appendix 1 for a brief history of this company) early in the year 1920. Later the same year, all the assets of NESCO were sold off to the Radio Corporation of America (RCA). Fessenden is also credited, along with Lee de Forest (the inventor of the basic triode valve), with the development of the technology where multiple sources of origin of audio signals could transmit at the same time. This technology was later to be termed as "amplitude modulation" (AM).

In the year 1915, David Sarnoff then still with the Marconi Company, was able to demonstrate in New York the transmission of music using radio technology. He then submitted a proposal to the company's management to take up commercial manufacturing of such radio boxes but his proposal was turned down purportedly with the following words: "The radio music box has no imaginable commercial value. Who would pay for a message sent to nobody in particular?"[22] Talk of poor decisions that changed electronics history!

So now, we had the technology of radio signal broadcast and reception in place. All that was needed was someone to actually start regular broadcasting to the public at large. We, of course, know of Fessenden's operation out of the Boston area, but this was only sporadic broadcasting.

History is not clear where formal broadcasting first began, largely because one or two of the broadcasters were not formally "licensed." A search through records indicates that the first regular radio broadcasts were started in Rock Island, Illinois, in 1907 with the station subsequently shifting to Iowa. However, there are many who say that this station was just an experimental one and that credit for starting regular scheduled radio broadcasting must rightly go to Charles Herrold.[23]

Herrold, an electrical engineer, had set up a College of Wireless Engineering in San Jose, California, and started broadcasting from there as early as 1909. His station received a formal license only in 1921. It has now morphed into the well-known station, KCBS in San Francisco. In October 1920, Westinghouse in Pittsburgh received the very first license for a commercial broadcasting station with the station designated as KDKA.

At the time the first broadcasting stations went on air in the United States, the first radio receivers were being made based on "crystal" technology. We have noted earlier in this chapter that Jagadish Chandra Bose had received a patent in 1904 for a radio signal detector using galena. In 1906, an American scientist, Greenleaf Whittaker Pickard, refined this further by attaching a fine wire termed a "cat's whisker," to galena and found that the signal reception quality was greatly enhanced. This was then the origin of the earliest and simplest of all radio receivers, which could usually be made at home using the improved galena detector, an antenna wire, a tuning coil, and earphones for listening to the audio signals without the need for batteries or a source of power.[24]

With Westinghouse's growing interest in the radio business, the company bought out de Forest's patent for AM and their engineers made significant changes to de Forest and Eric Tigerstedt's designs of the vacuum tube. Also at this time, Westinghouse bought out the patent of another American inventor, Edward Howard Armstrong (please see below), who in 1918 received the patent for what was described as the superhetrodyne receiver. The use of multiple vacuum tubes and the use of different frequency stages in this receiver enabled substantially enhanced radio reception. This system would then become the operating standard for radios around the globe.

In 1907, Arthur Atwater Kent, a graduate of the Worcester Polytechnic Institute in Massachusetts set up a company in Philadelphia called the Atwater Kent Manufacturing Works for producing a range of electrical products including contact ignition systems for automobiles. Kent sensed a future in the promising radio business and using his profits from the successful automotive business he put together a team to design and manufacture radios for the mass market.[25] By 1922, this company had put on the US market arguably the very first commercially produced radio set, the "Model 1" for the mass market.

Unfortunately for Kent, Westinghouse had already just a few months before introduced their own radio set, the "Aeriola" SR Receiver into the market and shortly thereafter the immensely popular "Radiola" range of sets, which were marketed by the RCA, the joint venture set up by GE, AT&T, and Westinghouse. Also by 1920, GE had established the manufacturing of radio sets but principally for marketing by RCA (see Appendix 2 for a brief history of this company).

By 1923, the radio market in the United States was beginning to see a boom with more broadcasters getting licenses and putting on air good and interesting programs. With that, the numbers of companies setting up manufacturing increased manifold. One report suggests that by the mid-1920s there were about a thousand manufacturers ranging from well-organized manufacturers down to small hobby-shop types. Of course, many perished over a period of time. Some of the more prominent names of manufacturers of the early 1920s were Federal Telephone & Telegraph Company, Crosley Radio, R. E. Thompson Manufacturing Company, American Radio and Research Corporation (AMRAD), American Auto and Radio Manufacturing Company, Grigsby-Grunow Company (brand name "Majestic"), Kemper Radio Laboratories, Remloc Radio Systems, and A. H. Grebe & Company. Many of these manufacturers were members of a group called the "Independent Radio Manufacturers," which had innovated and used some new radio circuitry, termed the "Neutrodyne," so as not to contravene the "superhetrodyne" patents by the grouping of GE, AT&T, Westinghouse, and RCA.

Of all the above radio manufacturers, it is Crosley that has the most interesting history. In 1920, quite upset with the price of $130 quoted to him for a radio for his son, Powel Crosley, from Cincinnati, decided to set up radio manufacturing on his own. Later that year, he came up with a model that would cost under $35. Before the end of the year, Crosley had bought out a company called Precision Electric and changed its name to Crosley Radio Company and began the first-ever manufacturing of low-cost radios. The first model, the "Harko" sold for just $7. Crosley was now dubbed the "Henry Ford of the Radio business." He would go on to introduce the first-ever car radio called the "Roamio" in 1930 as well as later introduce some rather uniquely designed refrigerators.[26] The company still exists and operates out of Louisville, Kentucky, and makes a whole range of retro radios, jukeboxes, turntables, and telephones, albeit under different ownership. Crosley's subsequently sold off the company to the Aviation Corporation (AVCO), which on closure of its own business sold the brand name to a group of investors.

By 1924, there were other and some bigger names that emerged as radio manufacturers. The most notable among these was Chicago Radio Labs, which was later renamed as Zenith Radio Corporation. Others included Philco (earlier called the Philadelphia Storage Battery Co.), Magnavox, and Sylvania.

Of these, Philco (see Appendix 1 for a brief history of this company), has had a most colorful history with ownership passing to the Ford Motor Company (renamed Philco Ford), then to GTE and finally becoming a part of the Dutch multinational, Philips. Sylvania, earlier a component supplier to Philco, subsequently became a partner and also ended up in the GTE basket and finally as a part of Philips. Magnavox, another of the great innovative US companies, becoming more famous for its phonographs, also ended up in the basket of Philips.

Zenith made history in 1924 by making the very first portable radio in the world. It would go on to manufacture some of the finest multi-band radios with superlative performance. Zenith in later years would also make a significant contribution in an aspect of television technology about which we will read in chapter 4. Zenith (see Appendix 1 for a brief history of this company), finally became a part of the Korean multinational LG Electronics.

Shortly after the radio craze hit the United States, it also arrived in Europe. In October 1922, the British Broadcasting Corporation (BBC) had been set up in Britain by a partnership between the Marconi Company, General Electric Co. (UK), the Radio Communications Company, Metropolitan Vickers, Western Electric, and the British Thompson-Houston Company. BBC began its radio broadcasts in November the same year from "Marconi House" located at the Strand in London.

William George Pye, earlier an engineer with the Cavendish Laboratories had set up a company, W. G. Pye and Company, to manufacture scientific instruments. With the start of radio broadcasting, the company decided to make radio sets as well and had its first product out just as BBC started broadcasts. Unfortunately, the first radio set was not a success. By 1924, with the induction of William Pye's son into the company, they developed their first best seller, the Model 700.[27] The company would go on to successfully make and market huge numbers of radio sets and also to become a pioneer later in televisions. It was in 1976 to be another company to be bought out by Philips. In the United Kingdom, Pye was to be followed into the radio business by E. K. Cole

(EKCO) in 1922, A. C. Cossor & Company (in 1927; also see chap. 4), Murphy (in 1928), and by Bush Radios (in 1932).

On the European continent, activity in the radio sector was also growing rapidly. Telefunken in Germany, which was already actively involved in radiotelegraphy equipment, started radio production in 1923 and by 1925 set up radio broadcasting as well. The Ideal Company (renamed as Blaupunkt) and RadioFrequenz Gmbh (subsequently named Loewe) in 1923, and Grundig in 1930, followed thereafter. Both the latter companies were subsequently acquired by Philips. Radio production companies also sprang up in Austria (Radiola and Berliner), Switzerland (Aluphon, Zellweger, Thorens, and Carma), Sweden (Baltic), France (Familial and SNR), Italy (Allochio Bacchini), Finland (ASA), and Norway (Elektrisk Bureau A/S) among others. Even in far-off Australia, a joint venture between Marconi and Telefunken was set up and was called Amalgamated Wireless Australasia.[28] The age of the radio had now well and truly arrived!

By the late 1920s, automobiles began proliferating in the United States, thanks to the mass production techniques introduced by Henry Ford. Also by then, subsequent to the Federal Highway Act of 1921, and the highway and parkway boom that followed in the decade thereafter, Americans were driving considerably longer distances over, by the then standards of the day, good roads and some onboard entertainment was required for the driver and passengers.

Some efforts had been made in the 1920s to use portable radios in cars. Significant among these were efforts made by the All American Mohawk Corporation (the "Lync" radio) and by Radio Auto Distributors (the "Airtone 3D" radio) around 1926.[29] By 1929, the Automobile Radio Company (later to be acquired by Philco) and Delco (later to be part of General Motors) also had a few automobile radios in the market with the brand names of "Transitone" and "Automotive 3002," respectively.[30]

In 1930, the Galvin Manufacturing Company, started by Paul V. Galvin and his brother in Chicago in the year 1928 to manufacture battery eliminators, decided that what the automobiles needed was a truly high-quality radio so that the driver and passengers could get over the tedium of driving over long distances as also get the news and weather information as they drove along. In 1930, this company introduced what was then called the "Motorola Radio" arguably the first-ever mass-produced car radio in the world. Shortly thereafter, the company introduced a line of radio sets for the police department and went on to make a range of highly successful two-way communication sets ("walkie-talkies") all using the Motorola brand name. By the 1940s, as often happens, the brand name became much more popular than the actual name of the company. So much so that, in 1947, the name of the company was officially changed to Motorola Inc. (see Appendix 1 for a brief history of this company).[31]

Of course, in later years, there were other very significant developments in radio technology such as the introduction of frequency modulation (FM) to give clearer reception over shorter distances, developed by Edwin Howard Armstrong, a graduate from Columbia University, New York, who also developed the superhetrodyne receiver. History has it that Armstrong eloped with Sarnoff's secretary and got married, yet subsequently he did work with RCA for a short time developing technology for noise reduction in radio broadcasting.

Armstrong showed in 1933 that by increasing the bandwidth rather than conserving it, as was at that time the norm, one could actually substantially reduce "noise" or "static" in radio broadcasts and obtained a patent for this.[32] He also then personally financed and set up an FM radio broadcasting station, W2XMN, in New Jersey. Sarnoff and RCA saw a clear threat from this technology to their own burgeoning radio business and somehow managed to get the Federal Communications Commission to allocate the frequency band that was being used by Armstrong for community radio, thus more or less finishing Armstrong's own plans for commercial exploitation of the FM radio business. A long unsuccessful legal battle ensued not only with RCA but also with De Forest. By the 1950s, Armstrong was now more or less financially ruined, his marriage broken, and in 1954 tragically this great inventor committed suicide by jumping down from his apartment in New York.[33]

One of the most significant developments in radio technology would be the miniaturization of radio sets using solid-state technology. Conventional radio sets being large could not be carried around for personal use. By 1947, the "transistor" had been invented (we will read about this in chap. 9 of this book). The small size and power requirement of this component meant that associated circuitry could be easily reduced and a full working set in a case could potentially be light enough to be carried around.

Several companies from around the world were trying to be the very first to make a small, portable radio set. The first to get there was a company called Regency Electronics with its model TR-1, in October 1954 in collaboration with Texas Instruments (TI). They beat the two other main contenders, Raytheon and Tokyo Tsushin Kogyo (later to be Sony) by about a year.

In September 1945, two former RCA engineers, John Pies and Joe Weaver, decided to set up their own company, Industrial Development

Engineering Associates, in Indianapolis, to provide consulting services as also to manufacture items like voltage stabilizers. They expanded their product line to make signal boosters for Sears Roebuck & Company, in 1950 and since electronics was going to be their "big thing," the name of the company was changed to Regency Electronics.

At about this time, one of the leading transistor manufacturers, TI, was looking for an electronics manufacturer to partner. One who would use their transistors in large quantities. The bigger companies like Philco, Sylvania, and RCA were for some odd reason not interested but Regency saw a great opportunity in collaborating with TI and in a short time, by October 1954, had developed a good working transistor radio only about 16 cubic centimeters in volume. Regency received a patent for this product.[34] In 1989, the name of the company was changed again to RELM Communications Inc. It is still in operation but now manufactures mobile communication equipment, based in Florida.

Tokyo Tsushin Kogyo, the Japanese company competing to be the first one out with a small transistor radio, was set up in Tokyo in the year 1946 by Akio Morita and Masaru Ibuka. In 1950, the company came out with a magnetized-paper recording tape and shortly thereafter a magnetic tape recorder. They started to work on a transistor radio in 1954 and by 1955 had produced an all transistor radio. Although in size this was smaller than the TR-1 transistor radio made by Regency, available records indicate that Regency was the first one to make it. With the success of their tape recorder as well as of their transistor radio, the name of the company was changed to Sony in 1955.[35]

Many other manufacturers would then follow these two leaders. These include familiar names such as Zenith, RCA, GE, Admiral, Magnavox, Motorola, and many others. In a short space of time, the transistor radio would become a ubiquitous consumer item with manufacturers from Japan and other Far Eastern countries producing multiple models at very low prices, more or less dominating the global market for many years.

Internet and Satellite Radio

In 1976, the United States passed the Current Copyright Act, which in its latest form applied to radio broadcasting using newer technologies. In 1998, President Clinton signed into law the Digital Millennium Copyright Act, which laid out specific rules for Internet broadcasts. This was particularly of interest for the then nascent field of the Internet (see chap. 8). The first entity to get into Internet radio was in 1993 when Carl Malamud set up the Multi-Casting Company of Washington.

In 1994, the student's radio station at the University of North Carolina at Chapel Hill had started a round the clock Internet "simulcast" of their terrestrial broadcast. In 1995, Robert Glaser, a former senior manager at Microsoft set up a company by the name of Real Audio, which produced software that enabled Internet radio broadcasts at higher speeds and better quality. It also made it possible for listeners to obtain earlier and archived broadcasts.

In 1995, Mark Cuban and Todd Wagner set up a company, Broadcast. com to do Internet radio broadcasting from a friend's computer in his bedroom. Shortly thereafter in 1996, Britain's Virgin Radio would become the first broadcaster in Europe to put all its radio broadcasts on the Internet. The well-known Internet company, Yahoo bought out Broadcast.com in 1999 for the sum of $5 million.

Today most of the world's major radio broadcasters have programs live, time delayed, or archived on the Internet. The popularity of this form of radio has grown by leaps and bounds. Today, estimates are that about a hundred million listeners get their radio feed from the Internet. The numbers of stations with live webcasts and streaming of music, news, and other content, runs into many thousands now.

In 1992, the United States Federal Communications Commission dedicated some radio frequencies for broadcasts using satellite. Subsequently, an auction took place in which two companies were held successful. These were American Mobile Radio (subsequently named XM Radio) and CD Radio (subsequently named Sirius).

Another company, Worldspace was set up with great fanfare to cater to the global market, with uplinking to a satellite from Maryland. The company evoked great interest as it was set up by Noah Samara, an Ethiopian expatriate in the United States. It streamed different genres of music through different channels 24 hours a day. Sadly though, despite having several hundreds of thousands of paying subscribers, the company soon ran into major financial problems. It shut down operations in 2010. Some ex-employees have now bought out the company but the service will no longer use the old dish antennae receivers. It will now be available only through mobile phones, direct-to-home satellite television channels and, believe it or not, the Internet!

CHAPTER 4

Television*

Television enables you to be entertained by people you wouldn't have in your home!

—David Frost

Even while the really exciting developments were taking place in telephony as well as radiotechnology, a bunch of very determined scientists and inventors were working feverishly to try to transmit pictures onto a screen. Many contraptions were tried but without much success. The origins of being able to transmit "pictures" perhaps may be found in the experiments in 1856 by Abbe Giovanna Caselli, an ordained Italian priest who also had a great interest in physics, electricity, and magnetism. Caselli was able to demonstrate the transmission of a single image over a wire using a machine he called the "pantelegraph."

This machine may perhaps be described as a sort of very basic facsimile (fax) machine as we currently know. Anyhow, by 1858, Caselli was able to develop the machine further and in 1860 was able to get Napoleon to finance the usage of the pantelegraph machines in the French telegraph network. Significantly, Caselli received a European patent for his device in the year 1861 and a US patent the following year, 1862, a full 14 years before Alexander Graham Bell formally received his US patent for the telephone.[1]

In 1873, an Irish telegraph operator, Joseph May, noticed something quite unusual. He found that when bars of selenium were exposed to the rays of the sun, their resistivity changed. This implied that changing intensities of light had the potential of being converted into electric signals. The same year a British electrical engineer, Willoughby Smith, the chief electrician for the Telegraph Construction and Maintenance Company, UK, could show that when light was passed through glasses of different colors, the variations in the resistivity of selenium were different. This was proof enough of the photoconductivity of selenium.

George Carey, a surveyor and part-time inventor, from Boston (in 1875) and Constantin Senlecq from France (in 1881) then worked further on the photoconductivity of the selenium principle to try to make a practical device. They used very large arrays of lights on an image and a corresponding array of selenium photocells separated by a multiple rotating switch system. As the switches turned, they could let light from each source of light come on to the corresponding photocell. As one can easily visualize, this was quite an impractical device even though it laid the basis of what was to ensue.

However, by 1877, Carey had refined his invention into a more practical device, which he called the "telectroscope," a basic "television" camera, dubbed the "selenium camera," and also further developed the use of multicircuit mosaics constituting what very belatedly were credited to Carey as the first concepts of a complete television system.[2] Carey was later to go on to invent something called the "thermofone," which may be regarded as the harbinger of what today is called the mobile "cell phone." Most unfortunately, there are no records to show that Carey actually promoted his patents or set up any company to exploit them.

In 1880, two inventors, William Sawyer of the United States and Maurice LeBlanc of France worked on actual image scanning. LeBlanc's work laid the basic foundation for the modern development of television. He clearly outlined the five basic items required, namely, (1) a transducer for converting light into signals, (2) a systematic scanning of a picture or an object, (3) synchronization of the transmitter and receiver, (4) converting the signals back into light, and (5) a receiver screen for viewing the image.[3]

Meanwhile, Alexander Graham Bell, at his Volta Laboratories in Washington, DC, was also working on utilizing light rays for communications utilizing the resistance varying properties of selenium. In 1880, Bell demonstrated the transmission of audio signals using light but only over a limited distance. Bell called his invention the "photophone" for which he received a patent. Unfortunately, this device had serious limitations especially the effect of external interferences such as rain, clouds, and so on and was never a real success, although in later years it would be perceived as a harbinger of fiber-optic-based communications.[4]

A British barrister, Shelford Bidwell, and another of those with a great interest as a part-time inventor in the technology of electronics, in 1880, worked further on Bell's photophone and managed to transmit a simple photocell-scanned image over a short distance. Bidwell termed this process as "telephotography." Unfortunately, he was never able to achieve proper synchronization between the transmitter and the receiver.

It was, however, in 1884 that Paul Nipkow, a young student from Poland (then part of Germany), who came out with the concept of a rotating disc (the Nipkow disc) with perforations made in a spiral pattern that would enable a picture to be deconstructed and divided into a pattern of lines and dots. He received a patent for this in 1885. Nipkow himself did not pursue the commercialization of his patent but Germany's first television broadcast channel in 1935 was named after him in recognition of his major contribution to the technology.[5]

In 1869 Johann Hittorf, a German scientist had noticed that in vacuum tubes, charged particles coming off the negative electrode, would give off a flicker of light when they struck the glass walls. In 1887, another German scientist (like Nipkow, also from Poland), Eugen Goldstein, did further work on this phenomenon and found some important aspects, which some years later, would enable the identification of the phenomenon being caused by negatively charged "electrons." He termed these charged particles as "kathodenstralen" translated in English as "cathode rays."[6]

And then, much like Popov in the development of radio, along came another Russian professor from the University of St. Petersburg, and invented what can best be described as a sort of mechanical television. Boris Lvovich Rosing received a patent in 1907 for the "electrical transmission of images." Essentially this system used Nipkow's rotating mirror, which facilitated images from a camera through an array of photocells.[7] Tragically, Rosing was later, in 1931, exiled by Stalin to Arkhangelsk, where he died two years later.

While Rosing's system was essentially mechanical in nature as far as the actual signal generation was concerned, the very first to propose a fully electronic television system was a Scottish electrical engineer, A. A. Campbell-Swinton, who had already done pioneering work on radiography in medicine. In 1908, Campbell-Swinton proposed a system in which cathode ray tubes would be used as both a transmitter as well as a receiver. In 1911, he developed this concept further and proposed that the larger flat end of the receiving cathode ray tube have a light-emitting phosphor coating (similar to what Karl Braun had used for his oscilloscope). The end of the tube with the phosphor coating would appropriately glow on receiving the stream of electrons in a pattern of lines and rows, much like Nipkow's disc. Unfortunately, Campbell-Swinton's concepts remained just so, as neither he nor at that time, anyone else, could make a practical system using these principles.

In 1922, a 14-year-old student in Utah decided that his forte was working with electrical products and decided to try his hand at being a part-time inventor in electronics, having been greatly impressed with what telephony had achieved by then. It is said that one evening Philo T. Farnsworth read about the possibilities of simultaneously transmitting pictures and sound into homes much like what had become possible for voice audio over radios. Farnsworth, a young teenager, came up with the concept of rapidly transmitting pictures as individual lines of electron beams, and then magnetically deflect each line at a time together forming a moving picture.[8] This as we know today is a basic television "raster scan."

Unfortunately, with the death of his father, Farnsworth, then still a teenager, had to look for part-time work to support the family. This is when he met with some of his employers who were quite impressed with his concepts for television. They decided to back him financially in establishing the Crocker Research Laboratories. Farnsworth then moved to Los Angeles and set up a workshop in his living room. At the age of 21, Farnsworth received a patent for an "image dissector," in essence the first-ever television camera tube. Later the same year, he was able to demonstrate a full-working television system, using a chemistry experiment flask as the basis for a receiver tube.[9] Philo Farnsworth then converted Crocker Research Laboratories into first the Television Laboratories Inc., and then to Farnsworth Television Corporation (see Appendix 1 for a brief history of this company) in 1929, one of the world's first companies dedicated to television technology. In August 1930, Philo Farnsworth received a formal patent for the television system.

At the University of St. Petersburg in Russia, Boris Rosing had a rather clever student by the name of Vladimir Kosmich Zworykin who would help Rosing with many of his experiments and work on his version of television. After graduation in 1912, Zworykin was able to land a job at the Russian subsidiary of Marconi's company. However, shortly after the Russian revolution and the civil war, Zworykin left for the United States in 1918 where he managed to get employment with the Westinghouse Electric Corporation. At this company, in the 1920s he worked toward developing television systems based on Campbell-Swinton's concepts of using cathode ray tubes both for transmission as well as reception and received patents for his work.

Unfortunately, Zworykin was unable to convince the management at Westinghouse about the practical uses of his work. However, in 1929, Zworykin was indeed able to publicly demonstrate a more sophisticated cathode ray tube, which he called the "kinescope" as well as an all-electronic camera tube, which he termed as the "iconoscope." By then, in 1930, David Sarnoff the erstwhile employee of Marconi International Marine Company, and the one involved with signals from the ill-fated *Titanic*, had risen to a very senior position at RCA. Sarnoff hired Zworykin to come and head the then fledgling television department at RCA's Electronics Research Laboratories.[10] By 1934, Zworykin's television system was ready for commercial use.[11]

At about the same time that Zworykin was working on his concepts for television at Westinghouse, another young part-time inventor in the United States, Charles Francis Jenkins, was also working independently on developing a practical television system at his Charles Jenkins Laboratories in Washington, DC. Jenkins had already by 1896 been involved in developing a type of movie projector and in 1913 published a paper titled, "Motion Pictures by Wireless." It was only by 1923 that Jenkins was publicly able to demonstrate a working system.

On receipt of a patent and also the very first license for commercial broadcasting (station W3XK), Jenkins in 1928 incorporated the Jenkins Television Corporation, arguably the world's first-ever company, albeit only by a few days, devoted to television. Jenkins promoted his television station by selling some quite inexpensive television receivers based on the Nipkow disc principle. Jenkins never did receive much acclaim as his television systems were regarded as being of the mechanical variety as compared to the electronic systems of Farnsworth Television Corporation.[12] The company was sold off to Lee de Forest in 1931 and became De Forest Jenkins Corporation.

At the start of WWI, a Scottish student at the University of Glasgow, by the name of John Logie Baird, had his studies as well as his amateur work on photocells interrupted. However, immediately after the war, Baird returned to work on his interests in electronics, which by now had turned toward television. By 1924, Baird was able to demonstrate a very basic semimechanical working model of a television broadcast.[13]

By 1925, Baird was able to demonstrate a better working model at Selfridges department store in London, and in January 1926, he demonstrated in front of members of the Royal Institution, a further refined system, which he called the "televisor." The newspaper, *The Times* of London reported thus of that demonstration, "The image as transmitted was faint and often blurred, but substantiated a claim that through the 'Televisor', it is possible to transmit and reproduce instantly the details of movement, and such things as the play of expression on the face."[14]

By 1927, Baird's system was already being able to transmit over the 400 miles between London and Glasgow, whereas the best American systems were at that time capable of barely achieving the 200 miles between New York and Washington, DC. A report appearing in the publication *Popular Science*, however, described the basic Baird system as being made out of "derelict odds and ends—old bicycle sprockets, lenses from bicycle lamps, tin cans—mounted on a framework of old sugar boxes, and tied together with string and sealing wax!"[15]

Baird then set up in 1928 his own company, Baird Television Development Company Ltd., which would then make the first-ever transatlantic television signal transmission between London and Hartsdale, NY, as well as produce the first television program for the BBC in London. Unfortunately, the Baird system, being a semimechanical system had clear limitations and was rapidly getting obsolete. So much so that the BBC decided to stop using all Baird systems and opted for the more modern electronic systems, offered by Marconi-EMI, which had access to RCA's Zworykin patents.

In 1930, another company called Baird Television was formed. This company made television sets as well as equipment for television broadcasting. Some manufacturing was outsourced to Bush Radios but this was discontinued in 1939. The company would subsequently be acquired by Scophony Ltd., and finally ended up being owned by Radio Rentals Ltd.

Baird himself went on to do pioneering work on the development of color television, video recording, as well as on infrared-vision devices and photography ("Noctovision" as Baird described it). An interesting facet of Baird's work is that he had involved the great Finnish scientist, Eric Tigerstedt, the coinventor of the triode, about whom we have read in chapter 1 of this book.

By 1939, RCA had started experimental television broadcasts from the Empire State Building in New York. At the New York World Fair the same year, the company introduced its first line of television sets although some reports suggest that several of the sets had to be coupled with radios to get the audio part of the signal. RCA is also known to have manufactured at that time a limited range of television sets and a lot of these were exported, including to the then USSR.

However, the credit for introducing the first-ever commercially practical sets to the public at large in the United States, is given to the American scientist and inventor, Allen B. DuMont (a graduate from the Rensselaer Polytechnic Institute) who did so in 1924. DuMont had acquired a great interest in electronics and radio in his youth, and had by 1915 become the youngest person to acquire a radio operator's license, at the age of 14. After graduation, he joined the Westinghouse Lamp

Company working in the radio tubes production department, where reportedly the production increased from 50 tubes to 50,000 tubes per day under his management.

Lee de Forest the coinventor of the triode enticed DuMont out of Westinghouse into his company, the De Forest Company, and made him a vice president. By then, in 1931, De Forest had also purchased Jenkins Television Corporation. On being denied additional funding to improve the cathode ray tubes used in their television sets, Allen DuMont resigned and set up his own company, DuMont Laboratories to manufacture long-lasting, high-quality cathode ray tubes.[16] By 1938, the company had started manufacturing the very first all-electronic television sets in the United States.

Europe, in the meanwhile, had stolen a march on the United States in terms of assembling and selling television sets. In Britain the earliest-known sets were sold by the Baird Television Development Company as early as 1928 but these were basically radio sets supplemented by a neon tube behind a Nipkow disc. By 1936, a British Company, A. C. Cossor Ltd., had established the commercial manufacture of television sets. Cossor, established in 1859 to manufacture scientific glassware, was a principal manufacturer of radio valves and cathode ray tubes. It then diversified into manufacture of radios (see chap. 2) and television sets. During WWII, it became the supplier of choice for radar systems for the armed forces. In 1958, its radio and television business was sold off to Philips, and in 1961 the remaining part of the company was acquired by Raytheon of the United States.[17]

The Germans perhaps were the quickest off the mark in manufacturing television sets. In 1928, Telefunken had already demonstrated a television system and was given the honor of installing some systems for the broadcast of the Berlin Olympic Games in 1936. Toward the latter part of 1934, Telefunken started selling television receivers in the market albeit with limited capabilities. There are also reports that about the same time as Telefunken was starting up its television activities, another German company, Loewe Opta, was developing its own television system and television sets. Unfortunately with the coming to power of Adolf Hitler, the founders of the company, the Loewe brothers had to migrate to the United States because people of Jewish faith were being targeted by the Nazis.

Another interesting German company involved in the television field was Fernseh A. G., set up by John Logie Baird and Robert Bosch, along with the optics firm, Carl Zeiss. By 1932, the company had developed the first-known "outside broadcast van" for taking and transmitting pictures from outside of studios. The company also started production of television sets in a limited way in 1934. By the start of WWII, Robert Bosch had acquired complete control of this company and moved the company principally into television broadcast equipment.

The Japanese were not that much behind. They already had some experiments on televisions in universities as early as 1925. Actual manufacturing of television sets started in 1939 with *Nippon Electric Corporation* (NEC) and Toshiba being the very first to manufacture them. However, with the onset of WWII, this program came to a standstill. The 1940 Summer Olympic Games were scheduled to be held in Tokyo and massive television coverage was planned. However, because of the war, the games were shifted to Helsinki, Finland. However, after the war, the Japanese companies came strongly back so much so that their companies such as Sony, Panasonic, Sharp, and so on became world leaders in the field of color television set manufacture.

In the United States, several companies took up commercial manufacture of television sets. The first, according to available records, was the Daven Television Company from New Jersey in 1929. Their first set sold for some $75. In 1930, a little-known company from Boston, the Shortwave and Television Corporation produced an innovative set designed by their chief engineer, Hollis Baird (no relative of John Logie Baird of Britain). In 1928, GE introduced its "Octagon" television set and subsequently the "Globe."

Other companies such as Jenkins, Pioneer, and Western Television followed, as did RCA after the court ordered split with GE. There were other manufacturers, names now part of history. There was Andrea and DuMont in the mid-1930s. After WWII, there were Crosley, Airline, Hallicrafters, Emerson, Sentinel, and of course, Zenith, Admiral, and so on.

The years of WWII and for sometime after, were understandably grim, and entertainment was not on top of people's agendas. Furthermore, companies worldwide in the electronics field had been tasked with other priorities related to the war effort. Yet, the Columbia Broadcasting System (CBS), in the United States was pushing ahead with a significant development in television technology. In 1940, a team of CBS researchers under the leadership of Peter Goldmark came up with a mechanical television system based on Baird's work, which could show images in color. The quality, however, was not good, the sets were rather large and worse, were not compatible with the old black-and-white sets. Even an improved version displayed in 1946 did not meet with general approval.

Meanwhile, RCA had also initiated work on developing color television technology using a substantially different approach. They decided on using inventions of a French scientist, George Valensi, for a new color encoding system and that of a German scientist, Werner Flechsig, for what is called a "shadow mask." After a few years of competitive controversy, it was RCA's system that finally met with the approval of the National Television System Committee (NTSC) in 1950 and since then has been the de facto standard in the United States. A handful of the US television manufacturers then introduced their first color television sets in the market and the National Broadcasting Corporation (NBC) established by RCA in 1926 became the first major color television broadcaster in the United States.[18]

The other international color television systems were to emerge later, the SECAM system (Sequentiel Couleur a Memoire—French) in 1961 in France by the company CSF (later to become Thomson CSF and now part of the Thales Group) and the Phase Alternating Lines (PAL) system developed by Walter Bruch of Telefunken in 1962. The PAL system went on to become the system of choice in most of the world whereas the SECAM system (and a variation thereof) was prevalent in the Francophone countries and the USSR. The NTSC system remains uniquely North American to this day.

The trouble with the then prevalent television technologies was that the transmission and reception were confined to local geographies only. It took several days, for example, to record a television program, "aired" say in London, and for the taped broadcast to reach cities and stations in other countries. In short, there was no instantaneous international television, unlike what shortwave radio broadcasting had been able to achieve. There was also the problem of receiving terrestrial television signals in some remote and mountainous areas and a satisfactory solution was needed.

One particularly difficult hilly area was in the Schuylkill Mountains area of Pennsylvania in the United States, especially because of the ridge formed by the "Broad Mountain." An enterprising husband and wife team, John and Margaret Walson had set up in the 1940s the Service Electric Company for servicing and repairing of appliances manufactured by GE in the small town of Mahanoy in the Schuylkill area. In 1947, there were three television stations in the region so the Walsons' thought it worth their while to try to add to the sales of television sets to their business. In a very short time it was noted that their customers were unable to pick up proper television broadcast signals from any of the nearby stations.

John Walson came up with a somewhat, for that time, smart solution. He put up an antenna on top of a nearby hill and then connected this to his company premises as well as to the television sets of the company's customers by means of a "cable" and appropriate signal amplifiers. Thus was born the world's first Community Antenna Television (CATV) or as it is more popularly known today, "cable television."[19]

At about the same time that the Service Electric Company had established a CATV System, a former captain of the US Army Signal Corps, Milton Jerrold Schapp, decided to go into business for himself and in 1950 set up the Jerrold Electronics Corporation with an investment of $500 only. This company was to become a pioneer in using a black box (possibly the equivalent of today's set-top box) along with a cable feed from an antenna to provide good-quality television signals to receivers in remote communities. Jerrold Electronics was subsequently sold off to General Instruments.[20] Many years later, in 1971, Milton Schapp was elected governor of the state of Pennsylvania.

The Russians launched the world's first satellite, the "Sputnik," in 1957 signaling the start of the space race. It was, however, only some six years later that the United States launched the world's first communications satellite, called the "Syncom II," which facilitated the first satellite-based signal transmission between a US naval station and a US Navy ship off the coast of Nigeria, in West Africa.[21]

On July 23, 1962, the communications satellite "Telstar," designed by Bell Laboratories and built by AT&T, which had been launched from Cape Canaveral (now Cape Kennedy) some two weeks earlier, went live and communicated back to earth the first global black-and-white television images starting with that of the Statue of Liberty and a short 20-minute program hosted by the famous broadcaster, Walter Cronkite, starting with the now famous words, "Good evening Europe! This is the North American Continent live via AT&T's, 'Telstar.'" This transmission was reportedly viewed live by over hundred million in Europe.[22] The era of satellite television and instant live global news and entertainment had well and truly arrived.

By the 1950s and the early 1960s, the leading economies of the world had started to recover from the effects of the terrible WWII. People began to have somewhat larger disposable incomes, wanted to acquire some basic luxuries, and also needed good entertainment and news sitting at home. With television now a success around the world, the number of channels available for viewing started to proliferate, at least in the free world. But this "happy" situation brought in its wake a somewhat unique problem—the advent of the "couch potato."

Viewers who had settled down comfortably say with beverages or food in hand, on their loungers, on sofas, or in bed, were extremely inconvenienced by walking up to their television sets to change channels, increase or decrease the audio volume, or to skip the burgeoning and annoying commercials that were being increasingly aired! What was required was a device to remotely control television sets instead of, say, asking one's kids, as indeed many did, to walk across and press the requisite buttons or tweak the required controls on a set.

The Television Remote Control

The first-ever remotely operated "controller" for a television set was something called a "Telezoom" installed in one model of televisions made by the Garod Corporation (originally named as Gardner—Rodman Corporation) of Newark, New Jersey.[23] This was a small hand-held unit, with a press switch, but with a hardwired connection to the set, whose only function was to facilitate a small increase in size of the image on television.[24] We do know that Philco had introduced a sort of remote controller for their very top-of-the-range radios but these were of no use with television sets.

Thus, history tells us, to begin with, televisions started to be controlled "remotely," if one can apply the term, by running a cable wire connection from the set to a handheld contraption with buttons. The very first one, appropriately named, "Lazy Bones" was offered in 1950 by Zenith on its television sets. Even though Zenith touted this product as the ultimate in being able to control sets while just lazing around, the problems with the cable wire snaking around rooms leading to people tripping over or with the connecting wire being pulled frequently were obvious to all.

Eugene Polley from Chicago, freshly graduated from a local technical college, started his career at Zenith in 1935 as a stores clerk in the parts stocking department. During WWII, he was seconded to work with the US Defense Department on bomb fuses and radar systems, but after the war was over he came back to work in the television engineering department at Zenith. By 1955, he had invented the first-ever wireless remote controller for television sets, which he dubbed the "Flash Matic." The gadget, which looked like a standard torchlight, would, on pressing the appropriate button, beam a ray of light onto receptor photocells placed onto a television set and thus control functions for volume, change of channels, or turn the set on or off.[25] Polley is credited with having made the following statement about his pathbreaking invention: "The flush toilet may have been the most civilised invention ever devised, but the remote control is the next most important. It is almost as important as sex."[26]

Unfortunately, Polley's invention suffered from possible malfunction due to, for example, too much sunlight, or sudden changes in level of illumination in a room. Clearly, something better was needed. Another engineer then working with Zenith decided that he would substantially improve on the Flash Matic given its problems of sensitivity to light. Robert Adler, with a doctorate in physics, and another of the highly talented engineers and scientists who had fled from Germany and Austria during the Nazi rule, came to the United States, and found employment at Zenith. Adler designed his remote controller by replacing the use of light with sound waves that would set off the desired control function on the television set. Adler then refined his design to use ultrasound in the product that had been dubbed "Space Command." By the early 1980s, when Infrared rays replaced ultrasound, almost ten million television sets had been sold with remote controllers based on Adler's design.

CHAPTER 5

World War II: Radar, Sonar, Cryptography, and Beyond

Radar won the war; the Atom Bomb ended it.

—Dr. Lee du Bridge, Head, Radiation
Laboratory (MIT), WWII

On March 13, 1938, the Germans took over Austria and by September 1938 they had taken over a large part of then Czechoslovakia. Next, Germany attacked Poland on the September 1, 1939. Two days later, Britain and France declared war against Hitler's Germany, which set the scenario going for WWII.

By early May 1940, the German Army with their blitzkrieg Panzer advances had annexed the Netherlands and was rapidly advancing into Belgium. By May 10, the French forces along with their allies, the British Expeditionary Force, were committed to battle against the German Army led by Field Marshall von Manstein. The battle for France had well and truly begun. As history tells us, the German Army broke through the allied defenses in the Ardennes and raced forward toward the English Channel. The British Expeditionary Force retreated toward the French port of Dunkirk and for all intents and purposes, was comprehensively trapped!

Inexplicably, the German forces did not pursue their offensive at even half throttle at Dunkirk and also did not attempt a crossing of the English Channel although they had the full British Expeditionary Force totally cornered. With the help of every possible floating ship available to the British, along with other vessels, including river craft and boats, the world's greatest known military evacuation was carried out with over 300,000 troops making it back to the shores of Britain.

On June 18, 1940, Prime Minister Winston Churchill declared, "The Battle of France is now over and the Battle of Britain is about to begin."[1]

The German Air Force, the Luftwaffe, shortly thereafter started massive air raids over Britain in an attempt to get full air superiority prior to launching an all-out invasion across the channel. Only a miracle could now save Britain! And that came in the form of what we now know as radar. The British claim that radar, a term coined by the US Navy in 1940, saved the country from being invaded and defeated by the Germans in WWII is by all accounts undoubtedly correct and well founded, however, their other oft-repeated claim that the British were the real inventors of radar technology is, unfortunately, not quite true, as we will read below.

In the year 1904, a German researcher from Dusseldorf, Christian Huelsmeyer, came up with an antenna-transmitter-receiver combination that could detect metallic objects at some distance. He called this product the "telemobiloscope" and set up a company by the name of Telemobiloscop—Gesellschaft Huelsmeyer und Mannheim—to commercialize the invention for use on board ships as an anticollision device.[2] Unfortunately, when the product was offered to the German Navy, Admiral von Tirpitz was believed to have responded by saying, "Not interested! My people have better ideas." A somewhat tragic rejection as at a later date, if the company was properly backed and funded, their invention could have saved the good ship, *Titanic*.[3]

It was as late as 1933 that the German Navy under a program headed by their head of signals research, Dr. Rudolph Kunhold, restarted work on the Huelsmeyer invention.[4] The task of getting a fully operational system going was given to two enterprising engineers from Berlin, Paul Gunther Erbsloh and Hans-Karl von Willisen. In 1934, these two formed their company Gesellschaft fur Elektroakustische und Mechanische Apparate (GEMA), the first company to commercially produce electronic equipment to detect targets by reflections of radio waves, or the earliest-known version of radar. A version of the same device could also locate targets by underwater radio and sound waves or sonar as it is now known.[5]

Meanwhile, a brilliant engineer from the Technical University of Darmstadt, Dr. Hans E. Hollman, who had developed the world's first microwave telecommunication system, joined the team at GEMA. By 1935, GEMA had built a land-based radar system called "Freya" and a sea-based version called "Seetakt." Several of Dr. Hollman's patents were then transferred to the German company Telefunken who in 1936 built the high-performance "Wuerzburg" radar system of which a then commander of the German Air Force disparagingly said, "If you deploy this radar it will take all the fun out of flying."[6]

The two other major German electronics companies of the time, Siemens and Lorenz, also became actively involved in the German radar program. Lorenz, would become famous for making the "Lichtenstein radar," which was most effectively used by the Junkers aircraft during their raids over Britain in WWII.

The Lichtenstein radar made by Lorenz (the full name being C. Lorenz A. G.) has an interesting story related to WWII. In July 1944, a Junkers JU 88 night fighter aircraft landed at the British airfield of Woodbridge. It appears that the flying crew got disoriented in bad weather and somehow thought that they were bringing their aircraft safely down to land in the then Nazi occupied Netherlands. The technical teams from the British Royal Air Force as well as from the British scientific laboratories were flabbergasted to see the exceptional Lichtenstein SN-2 interceptor radar and the Flensburg as well as Naxos radar detectors on board this aircraft. That is the time, purportedly, the British first learnt of the incredible sophistication of German radar technology.[7]

The company C. Lorenz A. G., has, unfortunately, more to it in recorded history than just the above. Starting in 1880 with activities in telegraphy, the company subsequently also entered the fields of telephony and radio. In the year 1930, Lorenz was bought out by the German company Standard Elektrikizitatsgesellschaft, which although an autonomous entity, was actually a part of the US conglomerate International Telephone and Telegraph Company (IT&T).

During WWII, Lorenz was a very major supplier of electronic equipment, components, and radar systems to the German war effort. It even had a substantial shareholding in the famous German aircraft manufacturer Focke Wulf. Although technically owned by an American multinational, not only was Lorenz a strong supporter of the Nazi party but also allegedly used labor from the Jewish concentration labor camps.[8] So effectively, a US-owned company was in fact working for both sides in WWII much to the detriment of the efforts of the Allied forces!

After WWII, most of the factories and operations of Lorenz in the eastern part of Germany were taken over by the Soviets. However, many of the other operations were revived after the war and the company came back strongly under the name of Schaub Lorenz and was also actively involved in the production and sales of consumer electronic equipment. In 1958, IT&T reorganized its German operations under the name of Standard Elektrik Lorenz (SEL). By 1987, the company was sold off to the French conglomerate Alcatel. The consumer electronics part of the business was then sold to the Finnish company Nokia.

The company GEMA developed many specialized radar and other systems for the German Air Force, Luftwaffe. Toward the end of the war, the Soviets captured their factory but it is believed that the several thousands of employees at GEMA had by then managed to burn all their drawings and design data. It is further reported that the Soviets forcibly took away some key GEMA employees to Russia to work on their own radar development.[9]

Early German radar systems did, however, have a problem. They had to use the then commercially available vacuum tubes, but the very-high-voltage operations needed for operating the radar transmitters meant that the very large amounts of positive charges emanating from the positive (anode) side tended to cause substantial damage to the receiving negative side (cathode). The use of very robust tungsten cathodes partially solved the problem. Scientists at Lorenz came up with an improvement, the "oxide cathode tube." But even with these better performance tubes, power outputs obtained were limited thus restricting the range of radar systems equipped with such tubes.[10] Further, the oxide cathodes of that time tended to degrade quite rapidly seriously affecting tube and system performance.

In 1920, Albert Hull, a researcher at the GE Research Laboratory, while working on advanced forms of vacuum tubes invented a tube with a cylindrical-shaped coaxial anode and cathode that could produce a relatively higher amount of power. Hull termed this the "magnetron." Although the device was initially intended for applications in telecommunications, a derivative of the device was to have an extraordinary effect on the course and outcome of WWII. The magnetron, although giving better power performance than the oxide cathode tube tended to be erratic in its behavior. In 1934, A. L. Samuel of the Bell Telephone Laboratories, came up with an improved version by incorporating four cavities in the anode, the first "cavity magnetron." Even this device had problems especially those related to the cooling of the anode.

Scientists in several other countries were also working toward a superior cavity magnetron. The Japanese were trying to go forward from the work of Samuel. Two Soviet engineers, N. F. Alekseev and D. D. Malairov, published a paper on their development early in 1940. But possibly the first to get one really operational, albeit more by a bit of fluke, were two researchers at the University of Helsinki.

Vilho Raikkonen and Erkki Riipinen were working on microwave detector circuits in the fall of 1939. As part of their work, they needed to generate strong microwaves. The available magnetrons and klystrons did not suffice for their purpose so they decided to get a hybrid of sorts between the two. Without serious theoretical work and going primarily with a sort of gut instinct, the duo managed to get a device perfected and used in experiments on February 21, 1940. The core of their device, a thick copper cylinder had a tunnel bored through it and six smaller tunnels bored around it to provide a sort of "resonance." They called their device a "multi-cavity magnetron." This development was to have an enormously significant impact on, at that time, a fledgling electronics company in Finland called Nokia and on the development of some fantastic radar systems that enabled Finland to take on a hugely bigger enemy, the Soviet Union, and actually get the better of them during what the Finns call their "Winter War"!

At the start of WWII, two British scientists returned to the University of Birmingham to work on improving the magnetron for use in a British radar program, which they realized would be inevitably required as part of the war effort. By February 1940, John T. Randall and his erstwhile student and fellow researcher, Henry A. H. Root had successfully developed a high-performance, stable cavity magnetron, in a six-cavity design and where the anode was incorporated into the vacuum chamber and was water cooled. By May 1941, the two scientists had a working cavity magnetron that could deliver a huge megawatt of peak power.

This development would completely revamp radar designing and make the microwave radar a practical reality, just in time for their country to defend itself successfully in the Battle of Britain.[11] Of course, over the years this cavity magnetron would become much cheaper and versatile and would become the basic component in an appliance used today in every modern kitchen, the microwave oven (invented possibly by accident by Percy Spencer at Raytheon but first sold through the Amana Refrigeration Company, which it acquired in 1965).[12]

Unfortunately, Britain at that time did not have the manufacturing capability of large-scale manufacturing of these cavity magnetrons, hence a delegation led by scientist, Sir Henry Tizzard, brought the details, in great secrecy, to their allies, the United States requesting their help. The MIT Radiation Laboratories were set up precisely for performing this function and became one of the largest such programs of WWII although it was Raytheon that did the bulk of the manufacturing of cavity magnetrons under the program.

Of course, just having an essential component such as an effective cavity magnetron does not produce a good radar system. For Britain, the task of developing a really top-class radar system was given to a Scottish scientist, Dr. Robert Watson-Watt (later knighted). In 1933, he was working for the British National Physical Laboratory as the head of a department dealing with applications of radiotechnology when he was consulted by the then government about the possibility of the Germans possessing a sort of "death ray."

Having debunked this possibility, with a convincing written scientific argument, Dr. Watson-Watt in turn produced a report on "The Detection of Aircraft by Radio Methods." To convince the Royal Air Force as well as defense ministry officials, Dr. Watson-Watt built a system in February 1935 and successfully demonstrated it in utmost secrecy at a location in the North Sea.[13] With subsequent ample funding in 1939, Dr. Watson-Watt oversaw the design and installation of a large number of radar systems strung all along the southern and eastern coasts of Britain to provide timely information to the Royal Air Force to enable them to intercept incoming bombers of the German Luftwaffe. This was dubbed as the "Chain Home" system and history tells us, it was successful enough to prevent the German invasion of Britain.[14]

Prominent among the British companies involved in the WWII radar program were the then big electronics companies, Cossor, Pye, EMI, Plessey, and Decca. Cossor's military and professional electronics activities later became a part of the American company Raytheon. The company, Pye, was to become a part of Philips. By a strange twist of fate, Decca ended up as a part of the other major British electronics company, Plessey, which in 1990 was renamed as Plessey Siemens Radar. The two companies engaged in the radar programs of the enemies during WWII had now become a merged entity!

The British Chain Home radar installation was put in place by mid-1939 and rightfully has received acknowledgment for its phenomenal part in the Battle of Britain. But much like in the case of the cavity magnetron, Finland had managed to develop its own radar system at about the same time if not before the British. If one goes to the lovely town of Mikkeli, situated in the southern Savonia region of Finland, in addition to the lovely Saimaa Lake, one can also visit the WWII operational headquarters of General Mannerheim, the then operational commander and later to be head of state of Finland.

Mikkeli in 1939 was the site of the most advanced control room with information feeds from a string of radar stations covering most of the border facing the Soviet Union as also the Gulf of Finland. The radars for the Finns were made by Nokia inspired by the pioneering work of Eric Tigerstedt on advanced vacuum tubes, night vision systems, and electronic counter measures equipment.

By December 1937, the first successful trials were conducted on the Nokia ET 01 (ET standing for Eric Tigerstedt) with echoes recorded from a test aircraft. By 1938, the ET 02 had been developed with a receiving range in excess of 100 kilometers. By 1939, a substantially improved model, the NR (Nokia Radar) series, dubbed the "Kotkansilmissa" ("Eyes of the Eagle"), had been introduced and formed the backbone of the Finnish air defense system. As a result, the Soviet bombers flying into Finland received one of the highest kill rates from the Finnish Air Force, the "Ilmavoimat."

During the latter part of WWII, with the United States entering the war on the side of the allies and subsequent to that war, it was the large US companies that became the leaders in radar systems. The companies included Rockwell, Northrop Grumman, Lockheed Martin (later to be renamed as MEADS International) and of course, Raytheon, which as we have seen above took over the radar operations of Cossor. Also among the leaders was the French company Thomson-CSF. The professional electronics activities of the other major French companies, Dassault, Aerospatiale, and Alcatel were merged into Thomson-CSF. In the year 2000, this grouping was renamed as Thales.

As we write this chapter, new radar technologies are being developed. One promising new technology being experimented by Thales relies on a network of receptors to pick up broadcast signals from reasonably strong signals from television broadcasters in the area. The difference in the broadcast signal from that reflected by an aircraft flying within proximity is analyzed to give the coordinates of the aircraft. This system thus does not need large and expensive rotating antenna to send out strong signals in the area to determine the location of aircraft. Needless to say more, such innovative technologies are bound to follow.

Sonar

Those of us who have seen the movie about a Russian submarine, *Hunt for Red October*, may recall the now somewhat famous dialogue that went as follows:

> CAPT. MARKO RAMIUS: [acted by Sean Connery]: Re-verify our range to target—one *ping* only.
> CAPT. VASILY BORODIN: [acted by Sam Neill]: Captain I—I—I just...
> CAPT. MARKO RAMIUS: Give me a *ping* Vasily. One *ping* only please.
> (emphasis added)

The "ping" in the dialogue cited above refers to a sound signal pulse sent out under water by an electronic device (usually on a ship or submarine) to hopefully get a reflection back from a body or "target" within range, much in the manner that a radar system would do on ground or in the air. The device is called the sonar (Sound Navigation and Ranging), although the British also referred to it as ASDIC, which was derived from "Allied Submarine Detection Investigation Committee," a cover for the secret work that was being carried out during the WWI.

The tragic sinking of the *Titanic* in 1912 after it hit an iceberg highlighted the need to have the possibility of a nonvisual system that would be able to detect objects on or below the sea surface especially during bad weather or at night. The initial work on this was carried out by the Canadian scientist Reginald Fessenden (see chap. 3) in 1912 when he was working for the Submarine Signal Company, in Boston, Massachusetts. The company itself was established in 1901 and the initial focus was largely on the development of steam-operated underwater coastal warning alarms supplementing the role of lighthouses. By 1914, Fessenden was able to demonstrate the possibility of detecting icebergs at a distance of three kilometers deploying a sort of passive sonar system.[15] The Submarine Signal Company was merged into Raytheon in 1946.

With increasing use of U-Boat submarines by the Germans in the WWI, a much more efficient system was required. A joint program of the British and the French was then established, ASDIC, for developing much more sensitive detection systems. In 1918, the first such system under the program was developed by a French scientist, Paul Langevin, working along with a Canadian colleague, Robert Boyle with the device itself being called the ASDIC, an acronym for the program.[16]

By the start of WWII, the British had further refined this technology with different versions for surface ships and submarines. Many of the first effective sonar/ASDIC systems were manufactured by the British company, Plessey Naval Systems. In the 1980s, this company was taken over by GEC-Marconi to form Marconi Underwater Systems and through a few other iterations finally became Thales Underwater Systems.

It is reported[17] that with the entry of the United States into WWII, the British shared their sonar technology with the Americans. The United States developed an active sonar system (the "Herald" system) using underwater tripods and positioned these systems outside of their major ports to be able to give advance warning of any enemy submarine activity. The United States then also developed several high-performance ship borne radar systems, which proved their worth in the war against the Japanese—for the US Navy, the 51 submarines based at Pearl Harbor, Manila (Philippines), and the US West Coast[18] were the only ships that could retaliate against the Japanese navy after the attack on Pearl Harbor. Their newly fitted sonar systems came in most handy indeed.

Subsequent to WWII, sonar systems were modified for many applications in civilian life. Among these were searching for underwater wrecks, studies on undersea mammals, and not forgetting the several expeditions of Jacques Cousteau.

Cryptography

If one drives on the M1 motorway north out of London and after some 40 miles gets off at Junction 14 on the A509, one sees the sign posting for the town of Milton Keynes (in Buckinghamshire). A short distance further, there are signposts for "Bletchley." On a pretty little street there called "Sherwood Drive" stands what at first sight seems like a large country home of a squire with the grand name of "The Mansion." This is the famous "Bletchley Park," where during WWII some quite extraordinary and ultrasecretive work had gone on and which would come to have an extraordinarily profound influence on the field of electronics and "computing" in the years to follow.

It is only very recently that people have been allowed entry as also to talk about this most secret campus at Bletchley Park and what actually the activities were inside. The building site was at one time even listed for demolition, however, to the great relief of people far and wide, it has now been turned into a museum due to the great efforts of a bunch of enthusiasts, helped in great deal by funding raised by the efforts of the company Google and by one of its British-born executives, Simon Meacham.[19] Among other present-day activities at Bletchley Park is a Science and Innovation Centre where office space and other incubation facilities are available to early-stage, high-tech, start-up companies.

So what precisely was going on at Bletchley Park, also sometimes referred to as "Station X"? A group of the very best academic brains, mathematicians, engineers, and scientists in the country were positioned here along with excellent support staff, all divided into semiautonomous teams, allocated to different buildings.

Those who worked at Bletchley Park were sworn to lifelong secrecy during and after the war. They could not even discuss their work with people from another group or building. All possible resources were provided to them with approvals coming directly from Winston Churchill, the then prime minister, so much so that he ordered, "Action this day! Make sure they have all they want on extreme priority and report to me that this had been done."[20] Station X or Bletchley Park was the heart of Britain's determined efforts to break the German cipher system, the

"Enigma," which was the fundamental encoded communications and signaling system of the Germans in WWII.[21]

Of course, cryptography has been around for many millennia. David Kahn in his book on the subject[22] lists examples of cryptography as far back as 1900 BC, with the use of nonstandard hieroglyphs and by Julius Caesar's forces in 60–50 BC, by shifting the normal alphabet a few places. But it was only after the WWI that modern cryptography really came into its own.

In 1918, a German inventor by the name of Dr. Albert Scherbius, in collaboration with a Dutch engineer, Hugo Alexander Koch, developed an electrical cipher machine based on a system of moving rotor wheels. The German Army was not very interested. Scherbius then started the German company Gewerkschaft Securitas. This company produced the first machine based on Scherbius's design in 1920 and called it the "Enigma" machine. In 1923, Dr. Scherbius changed the name of the company to Chiffrier Maschinen Aktiengesellschaft for producing improved versions of Enigma. By 1925, the German Army and Navy had become the principal customers and modified the Enigma to their requirements and were confident that with these changes and with a daily change of encrypting key codes the Enigma transmissions could not be "broken," or in modern-day terminology, "hacked."

In 1928, however, an Enigma machine was intercepted by customs authorities in Poland en route to the German embassy in Warsaw. A group of first-rate scientists and mathematicians at the University of Poznan under the leadership of Marian Rejewski managed to crack the encryption of the commercial Enigma machine of that time. They then built an electromechanical simulator for the Enigma and called it "Bombe."

Just before the beginning of WWII, the Poles shared their knowledge of the Enigma with the British and the French and also gave them operational clones made in Poland. At the start of the war and the German occupation of Poland, the British Secret Service managed to spirit Rejewski and his associate, Henryk Zigalski, out of Poland. They were to be of invaluable help to British cryptographers and scientists. Unfortunately for the British, the Germans had changed the cryptography procedure for the Enigma at the beginning of the war. They were now altering the key codes multiple times a day instead of just once.

To complicate matters, it was then found that the Enigma was not the only encrypting machine that the Germans were using. For messages passed between the high command in Berlin and the most senior army commanders, the Germans used a machine called the "Lorenz" (code named "Tunny" by the British),[23] made by the company Lorenz about which we have read earlier in this chapter.

For administrative and security reasons, Bletchley Park and its activities (code named "Ultra") were placed under the charge of a very senior naval officer. However, the analytic, technical, and code-breaking activities were carried out by a spectrum of scholars, some just specialists in the classics or expert in solving the most complex crossword puzzles. The most senior cryptanalyst was Alfred Dillwyn (Dilly) Knox. Dilly Knox had already made a mark in breaking codes during the WWI and subsequently during the Spanish Civil War. It was Dilly that first broke through the Abwehr's (Germany's secret service) Enigma code and also was responsible for recruiting the one person, Alan Turing, who would go on to be recognized as the "mastermind" at Bletchley Park after Dilly sadly died of cancer in 1943.[24]

Alan Turing joined Bletchley Park at the beginning of WWII. Turing is given the credit for making the British version of Bombe, which he appropriately called "Victory" and which at peak was "breaking" some two coded messages every minute. It is believed that Turing personally was responsible for breaking the Enigma code used by the German U-Boat submarines, saving many lives and ships at sea.

In 1936, Turing published a paper titled, "On Computational Numbers," which foresaw the possibility of building practical stored program universal computing machines. After WWII, Turing joined the National Physical Laboratory, where he worked on the development of the Automatic Computing Engine (ACE), which was completed in 1950. Unfortunately, this great genius was to be literally hounded for his deviant sexual preference, was convicted in 1952, and actually faced the choice between imprisonment and chemical castration (which he chose).

Turing died in 1954 in circumstances that until this day have never been fully or satisfactorily explained but listed as poisoning by cyanide. In the year 2009, the then British prime minister, Gordon Brown, issued a posthumous apology to Alan Turing, as homosexuality had been decriminalized in the United Kingdom in 1967. In the year 2012, 11 famous people led by the famous scientist, Stephen Hawking, petitioned the British prime minister for a formal pardon to Alan Turing, a genuine national hero and arguably, the father of modern computing.

But during the later days of WWII, it was the Lorenz Tunny that really needed to be broken into! This task was beyond classical code breaking as also Turing's Bombe was of not of much use. A team was formed under the leadership of Turing's former teacher at University of Cambridge, Max Newman, a pioneer in computing. Assisting him and Turing were Thomas H. Flowers who had been working for the Telecommunications Research Centre of the British Post Office, Col.

John Titman a cryptographer from the British Army, and William Tutte from Trinity College, Cambridge, also joined the team.[25]

It was Flowers with his experience on working with vacuum-tube-based equipment who first proposed the concept of a really high-speed machine that would do the enormous data crunching. It was designed using electronic vacuum tube circuitry (some 200 valves) and having its own electronic power supply. The machine that the group developed was called the Colossus "computer" and so by January 1944 the world had its first operating programmable computer, at that time dedicated to cryptography, but would spawn a technology that would change this world in the years to come.[26]

The Colossus successfully cracked the Lorenz Tunny. More such computers were required and not trusting a private company with this ultrasecret machine, the production was taken up at a plant in Birmingham owned by the then British Post Office, today incorporated into British Telecom.[27]

After the war, Newman joined the University of Manchester and in June 1948 his Computing Machine Laboratory, working on Turing's principles had built the world's first stored program digital computer, the Manchester Mark 1, also fondly called the "Manchester Baby." Flowers became involved in developing an all-electronic telephone exchange and ended up working for IT&T. Sadly, none of those involved in developing Colossus was given their rightful due during their lifetimes as the work they had done was still covered under the Official Secrets Act. So secretive in fact that Winston Churchill said that it was "the Goose that laid the golden eggs, but never cackled."

Much like the "finest hour" for the Royal Air Force flying crew at the Battle of Britain, it was pretty much the same for British scientists having performed extraordinarily by developing, for that period what was cutting-edge technology, even given the exigencies of war. Regrettably, they were to surrender this lead shortly after WWII when other countries, especially the United States, in computing and Japan in consumer electronics would leave a postwar Britain, short on resources and shorn of its "empire," quite some distance behind.

CHAPTER 6

Computers and Computing

I think there is a world market for maybe five computers.
—Thomas J. Watson, Jr.

There is no reason anyone would want a computer in their home.
—Ken Olsen

The computer was born to solve problems that did not exist before.
—Bill Gates

What the computer is to me is the most remarkable tool that we have ever come up with. It's the equivalent of a bicycle for our minds.
—Steve Jobs

Thomas Watson Jr., Ken Olsen, Bill Gates, and Steve Jobs were all great men who headed great companies, namely, International Business Machines Corporation (IBM), Digital Equipment Corporation (DEC), Microsoft, and Apple but regrettably neither they nor their companies invented computers, as we will see below.

"Compute" derives from old Latin word with the meaning "to reckon." A standard definition in a dictionary is also that of calculation or reckoning. So if we go by these definitions then computing or "computers" as devices used as aids to calculation go way back into history. Through history, large calculations had always proposed a problem. After all, there were just ten digits on an individual's hand! So man was for long seeking a solution to this problem.

We do know, for example, that the Babylonians had a form of an abacus as early as 2400 BC. We also know that Leonardo da Vinci had developed a form of manual calculator by 1492. In 500 BC or so, the great scholar from India, Panini, had written the very first structured statements aiding in calculations—what today may be described as "software." By 300 BC, Pingala, another great mathematician from India had postulated the "binary system," setting the stage for the binary code that would become the basis for modern Boolean algebra and encoding of modern "computer" data.

Fast-forward to the year 1822, when Charles Babbage, a mathematician who failed to graduate from the University of Cambridge, designed his first mechanical computer, which he called the "difference engine." With many years of further toil, Babbage in 1834 came out with a considerably improved product, which he had now made somewhat programmable. He called this the "analytical engine."[1]

In 1880, the census count in the United States took forever to complete and it was feared that the next census count in 1890 may end up in a fiasco. In 1881, Herman Hollerith invented an electrically operated counting and tabulating machine based on the use of punched cards such as those used in the Jacquard weaving machines. With the success of his machine, in 1896 Hollerith established a company, the Tabulating Machine Company, to commercialize his invention.

Unfortunately, in 1905 Hollerith and his machines ran into problems with the US Census Bureau, who wanted the machine to be considerably improved. When Hollerith refused, the Census Bureau decided to make the machines themselves with the help some of Hollerith's former colleagues. What followed was protracted litigation and by 1912 Hollerith had lost the legal battle as well as most of his money.

Hollerith sold off his company in 1912 to the Calculating, Tabulating and Recording Company (CTR), one of the very first conglomerates in the United States.[2] Hollerith, however, continued as a shareholder and consultant. In 1914, a top-class salesman, Thomas J. Watson Sr., joined CTR. Watson then set about completely revamping the sales and marketing operations of CTR as also supervising the building of a much-improved product. By 1924, Thomas J. Watson was controlling the operations of the company (later to become its president) and had it renamed as International Business Machines Corporation (IBM).[3]

In 1908, an Irish accountant, Percy Ludgate, came up with a simpler version of Babbage's analytical engine. It was an electrically driven semi-portable computing device with a paper tape memory. Unfortunately, while there are references available to his technical paper on the device, there is no clear evidence that the machine was actually ever produced and satisfactorily demonstrated.

In 1925, a professor of electrical engineering from the Massachusetts Institute of Technology (MIT), Vannevar Bush, started work on the creation of "analog" computing devices for carrying out integration and differentiation calculus computation. By 1931, Vannevar Bush had finished the differential analyzer computer using a complicated system of gears. Subsequently, he completed an electron-tube-based differential analyzer computer. Interestingly, prior to his work on computing, Vannevar Bush in 1922 had set up a company called Metals and Controls Corporation, along with two friends, Laurence Marshall and Charles Smith to manufacture thermostats and other products including a Thermionic Tube that was called the S-Tube. The patent to this device was bought out by the radio manufacturer, AMRAD (see chap. 3). This company was later renamed as Raytheon in 1925.

Toward the later part of the 1920s, a student at the University of Iowa, the son of Bulgarian immigrants to the United States, John Atanasoff, had to write a dissertation that involved some pretty long calculations. He worked on a sort of clunky mechanical desk calculator for his calculations yet it took far too long and became tiresome since even simple multiplications and divisions required multiple additions and subtractions. He decided that there had to be a simpler calculator than those readily available at the time and started work on an electromechanical calculator of his own design. Although Atanasoff had the principles of his machine worked out by 1938, he could complete it only in 1940. It was called the "Atanasoff-Berry Computer" with the name of his assistant Clifford Berry added on.[4]

In 1936, a scientist working toward his doctorate degree at Harvard University on the theory of space charge conduction realized that the total number of calculations required to be carried out were perhaps simply not possible in the normal manner. Howard Aiken needed a suitable computer, but nothing was readily available. Aiken, greatly inspired by the earlier works of Babbage sought permission to build his own computer at Harvard but was denied permission. Aiken then approached the Monroe Calculating Machine Company with his proposal but was again turned down.

Aiken finally approached IBM. IBM agreed to build the machine but at a cost of some $200,000. Work on this giant of a machine dubbed the Mark I, weighing some five tons and comprising of three-quarters of a million parts, could only be completed by the end of 1943. IBM at that time did not have the expertise to make a fully electronic machine. The end result was a giant of an electromechanical tabulating computer.

Subsequently, Aiken was to design and make the Mark II, Mark III, and Mark IV machines at Harvard University. The Mark III was an electronic machine and used a drum memory whereas the Mark IV was an all-electronic machine with solid-state devices and a core memory. It is perhaps true to say that the Harvard machines really did not set any great trends or standards for the future of computers but it was Aiken's initiative in setting up the first graduate program of computer sciences at a university that had much more of an impact in computing. Several of the graduates from this program, American as well as foreigners, would go on to make great contributions at various computer companies such as IBM and Wang Laboratories (setup by An Wang, a student who had come from China).

Konrad Zuse worked as an engineer with Henschel Flugzzeukwerke, a German aircraft manufacturing company in Berlin at the start of WWII. He was another one of those who wanted to set about developing a machine to facilitate long calculations. By 1938, he had developed an efficient mechanical binary computer the "Z-1," in the kitchen of his parent's home. By the following year, Konrad Zuse had improved his device by making it electromechanical and incorporating a "keyboard" and some flashing lights to indicate results. This he called the "Z-2." By 1940, he had set up an entity called Zuse Apparatabeau, which was responsible for all the designing of his computers and is widely believed to have received some funding from the then Nazi government in Germany.

By 1941, Zuse had developed the "Z-3," a sort of electronic version, which was in some ways programmable, according to some reports, by the use of old cinematography films. However, he had no means of storing the program or data in the memory of the computer. Regrettably, for poor Zuse, the Nazi government then in power declined to provide any further finances for developing the next generation of his computer as they felt that would win WWII anyway. The Z-1, Z-2, and Z-3 computers and their drawings were destroyed during the war. Zuse, however, managed to smuggle out to Zurich, in neutral Switzerland, a basic version of the upgrade, the "Z-4" computer and managed to complete his work at the Federal Polytechnic there.

In the years that Zuse was working on the Z-4, he also started to develop what in later years would come to be called "computer software." This was in the form of the world's very first high-level computer language, which he called "Plankalkul." He then went on to demonstrate the use of this language in solving engineering and scientific problems.

By the year 1946, Zuse had established a company to commercialize his computers. This company, Ingenieur Hopferau, thus became arguably the world's first computer start-up company. It received some "venture capital" as part of a contract with IBM. In 1949, after WWII, Zuse returned to Germany and set up another company Zuse KG for the manufacturing and marketing of his now advanced computer products.

The company went on to manufacture computers with increasingly enhanced features including the "Z-23" in 1961, based on transistors, and finally the "Z-43" in 1964, which was based on modern ICs. Zuse KG was subsequently acquired by Siemens in 1966.

Two engineers, John Mauchly and J. Presper Eckert, from the Moore School of Electrical Engineering at the University of Pennsylvania, Philadelphia, had started work in July 1943 on a large computing machine that they called the Electronic Numerical Integrator and Computer, ENIAC for short. However, this digital computer, initially meant for calculating ballistic trajectories for the US Army, was only made fully operational and announced to the public at large in February 1946.[5] Eckert and Mauchly then set up their own company, at first called the Electronic Control Company with the name subsequently changed to Eckert-Mauchly Computer Corporation (EMCC).

EMCC was, however, to run into serious quality, delivery scheduling, and cost estimation problems as well as at one time security clearance problems during the Senator McCarthy probes into communist infiltration. This company then had to be sold off to Remington Rand in the year 1950. Remington Rand was later to be merged in 1955 with the Sperry Corporation to become Sperry Rand and finally, as it exists today as Unisys after a merger with Burroughs Corporation.

Remington Rand started their computer related activities in the year 1949 and introduced in the US market their Model 409, which they described as the first "business computer" although the specifications of the machine would not have qualified it to be called so under the computer architectural parameters that came to be widely accepted. This machine was subsequently sold as the UNIVAC 60 and 120, and was to become the first computer to be installed in Japan.

Burroughs Corporation had started life in 1888 as American Arithmometer Company, in St. Louis for the manufacture of mechanical calculating machines. The company was established by William Seward Burroughs, but unfortunately he passed away in 1898 from medical complications. In 1904, the company moved to Detroit and the name changed to Burroughs Adding Machine Corporation. Over the next few years, this company was to take over a few more manufacturers of calculating machines and by 1953, with its increasing interests in the computer business after the purchase in 1958 of Electrodata Corporation, it was further renamed, this time as Burroughs Corporation.

The University of Pennsylvania was not the only US university making great attempts at developing computer technology. At MIT, in Boston, a project had been initiated in 1944 to develop a high-speed flight training simulator for which very high-speed computations were required. The project leader Jay Forrester decided that for his computer to be able to deliver the required performance, a memory with very high levels of speed and reliability was required, which was not available by the use of electrostatic storage tubes, then in use for the "memory."

Forrester developed a new memory system that used a mesh of ferrite rings and metallic wire looped through them. This was the development of the magnetic core memory for which Forrester received a patent. By 1951, the MIT computer system, dubbed the Whirlwind was operational.[6] The Whirlwind was the fastest computer of its time and although using far fewer components than the ENIAC of the University of Pennsylvania, it was considerably more powerful and could do for the first time, real-time computations in microseconds.

One of those involved in reviewing the work on ENIAC and its planned successor EDVAC ("Electronic Discrete Variable Automatic Computer," subsequently called the UNIVAC, short for "Universal Automatic Computer") was a brilliant mathematician from the Institute of Advanced Studies, Princeton University, and a colleague of the great Albert Einstein. John von Neumann in a report written in the year 1945 for the first time spelt out in considerable detail what the basic structure of a modern computer should be.

Von Neumann proposed that all programs and data in a computer be stored together (main memory) and that as and when transactions are to be done only the required data and programs need be transferred to a considerably faster "memory" (random access memory, in current-day parlance). The processing of the data using the relevant programs would then be done by a core processor called the central processing unit (CPU, as we know it today). This, the "von Neumann architecture" was to become the established standard structure and architecture for all modern computers, mainframe, desktops, laptops, and so on.

On the basis of the von Neumann computer architecture, the first true modern computer would have to be the one developed at the University of Manchester in Britain. Max Newman one of the principals from Bletchley Park, joined the Computing Machine laboratory at the University of Manchester after WWII and by June 1948 had developed the world's first stored program computer, the Manchester Mark 1 (also known as the Manchester "Baby"). His colleague from Bletchley Park, the great Alan Turing, having joined the National Physical Laboratory after the war went on to develop there the computing engine, ACE.

A substantial part of the efforts to develop computers at the University of Manchester were supported by a British company called Ferranti,

which having been established in 1882, was in the business of transformers, electronic components, domestic appliances, and so on, and
had played an active part in delivering electronic systems for the war
effort in WWII. Ferranti had identified computers as a growth area.
In addition to the Manchester Mark 1 and Mark II developed with
the help of the University, Ferranti through the period starting from
the mid-1950s through to 1962 introduced the "Pegasus," "Mercury,"
"Perseus," "Sirius," and the "Orion" computer systems.[7]

In the year 1962, Ferranti introduced the fully solid-state "Atlas" system designed in collaboration with first, the University of Manchester
and later with the University of Cambridge and was at that time supposed to be the most powerful computer in the world possibly just ahead
of the "Stretch" system developed by IBM. Unfortunately, by 1963, with
growing local competition as well as from systems from IBM, Ferranti
decided to sell off its computer business to International Computers and
Tabulators (ICT).

ICT had started as the British Tabulating Machine Corporation,
which for years sold in the United Kingdom machines made by
Hollerith (later to be IBM). In subsequent years through a series of
mergers between ICT, Elliott Automation, and English Electric, the
company became International Computers Ltd. (ICL), and was finally
taken over by the Japanese company Fujitsu who in turn arranged for
ICL to take over the operations of Nokia Data from Finland.

If we note the tremendous contribution for the development of modern computers made by the University of Manchester team initially
set up by Max Newman based on the concepts of Alan Turing, for
all intents and purposes the real start of computers as we know them
today, undoubtedly was at Bletchley Park, England during WWII (see
chap. 5). The pioneering efforts of Alan Turing, Max Newman, and
especially those of Thomas Flowers, in making by December 1943, the
"Colossus" computer, of course, has been acknowledged and recognized
by one and all.

Ferranti, however, was not the only British company making computers by 1950. Elliott a company started in 1804, was making telegraphy equipment in the late 1800s and then during the WWI was making
some electromechanical equipment for the Royal Navy. During WWII,
they were given the task of producing secret radar systems and digital
gun control systems for the admiralty. Very little of this secret work was
ever known; however, it most likely involved digital computing techniques, as shortly after the war, by 1950, Elliott, had come out with its
Series 400 computers.

Between 1956 and 1967, Elliott had a marketing tie up with the American company National Cash Register (NCR) for their computers and by 1961, had a 50 percent market share of the computer market in the United Kingdom. However, because of financial problems the company first sold out to English Electric and subsequently was merged into ICL.[8]

NCR in the United States was founded in 1884 to manufacture mechanical cash registers. By the year 1906, it had made the first electric motor powered cash register. In 1952, it also got bitten by the "computer bug" and bought out a medium-sized company called Computer Research Corporation in Hawthorne, California, that specialized in making small digital systems for use in aviation. Using this newly available computer expertise, NCR produced its first transistorized business computer, the NCR 304 in 1957. In 1991, NCR was acquired by AT&T but in 1996 changed its name back to NCR before exiting the computer hardware business entirely and concentrating on software and solutions.[9]

When WWII started, a bright mathematician and radio amateur, Maurice Vincent Wilkes left his post at the University of Cambridge in England to work on the nation's Radar program where he picked up a good knowledge of electronics. At the end of the war, Wilkes returned to Cambridge and became the director of the mathematical laboratory there. By then, he had read about the work in the United States of computing pioneers, Eckert and Mauchly as also that of Aitken. Wilkes had also read a copy of von Neumann's report and put up a proposal to the university for a plan to catch up with the Americans in the development of modern computer technology.

Wilkes visited the Moore School of Electrical Engineering at the University of Pennsylvania to see developments in their newly started EDVAC program. Rather strangely, this program did not have a security cover at that time. In May 1949, Wilkes initiated the EDSAC (Electronic Delay Storage Automatic Calculator) program at the University of Cambridge, and in a short time had developed a fully functional stored program electronic computer.[10] Which came first—the Manchester Mark 1 or the Cambridge EDSAC? If one goes strictly by the dates of successful demonstration then clearly Manchester came first although it would appear that performancewise, the Cambridge EDSAC had the edge. In any case, the Cambridge effort spawned the greater commercial effort in British computing.

The prize for the most audacious effort at manufacture of computers must, however, go to a British company purveying, of all things, tea! J. Lyons & Company, had a very successful tea business with many outlets around the country. The company sort of did by themselves all things needed for their business with minimal outsourcing. This included tea estates, production, laboratory, packaging, transportation et al. After WWII, the company managers realized that with rapidly increasing costs including those for rentals of premises they needed to introduce efficiencies into their systems. In 1947, two of their managers on returning from a visit to the United States reported to their board of directors that the future lay in introducing at the earliest computers into their processes.[11]

Since computers at that time were not readily available commercially, J. Lyons & Company decided to give the University of Cambridge some money as sponsors for a project to assist Lyons in developing computers that the company could use. The university was at that time already working on developing their EDSAC systems under Maurice Wilkes. By 1951, a team put together by the company and assistance from the University of Cambridge, managed to put together a working computer, together with basic software. This was called the Lyons Electronic Office (LEO for short).[12] A company, LEO Computers Ltd. was set up to commercialize this development.

Over the next two years, LEO Computers produced improved versions, the LEO II and LEO III along with some high-speed printers. Many machines were sold to other British companies and several were exported. However, by the 1960s the company was faced with very stiff competition from much better performing systems from the United States, available at competitive prices. The company was sold to English Electric in 1963, and finally ended up becoming a part of Marconi.[13]

At the beginning of 1960, computers and computing started to be dominated by the United States. The main computer companies from the United States were at that time described as "Snow White and the Seven Dwarfs." The seven "dwarfs" were Burroughs, Control Data Corporation (CDC), GE, Honeywell (which had bought out Raytheon's shares in Datamatics), NCR, RCA, and UNIVAC. "Snow White," of course, was the emerging giant and the dominant player, IBM. GE and RCA were soon to exit the computer business with GE selling out to Honeywell. Honeywell itself would merge with the French combination of Compagnie Industrielle pour l'Informatique (CII) and Compagnie des Machines Bull along with the Japanese company Nippon Electric Corporation. By 1991, Honeywell had also exited the computer business completely. CDC after its main designer, Seymour Cray, left to start his own company (Cray Computer Corp.) for making "super computers," became primarily a computer peripherals company and in 1992 changed its name to Control Data Systems Inc.[14]

In 1953, IBM had a bright young engineering manager, Thomas Watson Jr., the son of the chief executive officer. He wanted to design a brand new machine that would address the needs of the US defense sector especially with the war in Korea reaching a high point. His father, Thomas Watson Sr., was not particularly keen on having a new product that would take away from IBM's profit making punch card processors, but gave in to his son's enthusiasm.

The result was the development of the IBM 701 computer system, which was built with an electrostatic tube memory, and used magnetic tape to store data. For IBM, this was the first commercially successful general-purpose computer although principally for use for scientific calculation. It also spawned the writing of the programming software language, FORTRAN (*For*mula *Trans*lation).

However, the slow speed of the IBM 701 at that time meant that a redesign was soon required. First, the IBM 608 advanced calculator was made in 1957 using some three thousand transistors, but continued using the punch card input/output data handling. Then, the IBM 704 computer was made using the ferrite core memory concept developed by Forrester. In 1956, Watson Sr., passed away and Thomas Watson Jr. became the new CEO. By 1960, IBM had made its first stored program, core memory computer using transistor components. This, the IBM 1401, was at that time the fastest commercially available computer in the world.

Carrying forward his plans for moving IBM into the forefront of computers in the world, Thomas Watson Jr., put in place in 1960 the development of IBM System 360, the first computer system to use widely available software and peripherals such as printers and disc memories. The term "IBM Compatible" had now come to become part of everyday computer talk!

The Americans and the British were not the only ones working furiously on computer technology in the early 1950s. According to information that became somewhat more readily available after the end of the Soviet Union and the demise of the "iron curtain," the Soviets had put in place a computer development program of their own under the leadership of a famous scientist from Kiev, Academician S. A. Lebedev. By 1950, the Soviets had also made a stored program computer, the Malaya Elektronno SchyotnayaMashina (MESM). In subsequent years, Lebedev was to oversee the development of improved computers with the Bolshaya Elektronno Schyotnaya Mashina (BESM) series machines made in Moscow.[15]

Meanwhile in the city of Minsk, another Soviet scientist, Isaak Brook, created the somewhat faster "M" series of computers also known as the "Minsk" series. Unfortunately all these computers, however good they may have been, were primarily focused on applications in the nuclear weapons and space programs of the Soviet Union using nonstandard (by Western standards) components, peripherals, and software. These computers were thus not readily available to the nonmilitary and the nonspace sectors.

With serious worries that the Soviet Union and its COMECON (Council for Mutual Economic Assistance) allies would be left far behind by the rapidly advancing computer technology in the West, they embarked on a program of "stealing," reverse engineering, and copying as much of Western computer technology as they could. Some reports suggest that the Russian KGB was actively involved in this game! Because of these activities, the Soviets were able to make copies of the IBM computer series. The ODRA computers made in Poland were based on ICL technology. Computer peripherals then available from Bulgaria looked suspiciously like copies of those from CDC.

Minicomputers

However, all the computers that we have referred to above, regardless of the "generation" (first, second, or third) were what were termed as "mainframes." These were large, mostly stand-alone machines churning vast amounts of data and requiring huge amounts of space and air conditioning with dust control. Above all, they cost a pretty packet, some as much as hundreds of thousands of dollars! Hardly the way to make computers and computing available to larger sections of society and enterprise! An opportunity existed that was just waiting to be addressed, especially as by then smaller and more reliable electronic components had started to become widely available, particularly in the form of semiconductors.

The very first to spot this opportunity were two young engineers who had worked on the Whirlwind project at the Digital Computer Laboratory at MIT. Ken Olsen and Harland Anderson received some venture capital funding in 1958 to set up a company, DEC, near Boston. They started on a small scale producing what may best be described as subsystems on a printed circuit board.

But by 1960, DEC had made its first full computer, the PDP-1, with PDP being the short form of "Program Data Processor." This device came along with a screen display to show what was going on. It was no larger than a common refrigerator and better still it cost a fraction when compared to the cost of the large mainframes of that time.

DEC made a few more computers until 1965 in which year they unveiled the PDP-8, widely regarded as the world's first true "mini-computer." It is believed that over 50,000 of these machines were sold. A new era in computing had now begun![16] By 1970, DEC introduced to the market a low-cost computer, the PDP-11, which was priced at $10,000. This new machine surpassed even the PDP-8 by selling well over 500,000 systems.

However, Harland Anderson decided to leave the company. Also some of the main engineers of DEC, in particular one of the designers of the PDP-8, Edson de Castro, left DEC to set up a competitor company, Data General. This resulted in delays in the further development of products by the company. It started losing market share and as a result resources became tight. By 1998, the company was sold off to Compaq, which in turn got acquired by Hewlett & Packard (HP) in 2002.

Data General, as has been noted, was a start-up company established by some of the engineers who had left DEC. The company, set up in 1968, about a decade after DEC had started business, wanted to build a low-cost, fast, multipurpose computer that would take away from DEC's commanding share of the market. With a somewhat aggressive style, the company introduced their NOVA series, which was a huge success and enabled the company to go public within a year of starting sales. Within a decade of commencing operations, Data General had become the fastest growing computer company and featured among the Fortune 500 companies.

The rapid growth of the company brought in its wake a host of problems. There were management issues, delays in new product development, software problems in its new line of smaller computers, and of course, growing competition as other manufacturers entered the field of mini-computers. IBM had also by then revamped its product program. Data General suffered continuous losses over a period of years. A joint venture with a Russian company was announced but did not really take shape. de Castro was asked to leave the company but by then damage had already been done. In 1999, EMC Corporation, a computer peripherals company, took over the company and its assets and much like DEC another great computer company had become only another name in history.

In 1935, two graduates in electrical engineering from Stanford University in California, who had studied under the famous Professor Frederic Terman, decided to go into business on their own. Bill Hewlett and Dave Packard established Hewlett & Packard (HP) in a garage (at 367, Addison Avenue, Palo Alto, California) to manufacture instruments for the burgeoning business of electronics. HP as a company was to start the trend of start-up high-technology companies in the Bay Area south of San Francisco, which later came to be popularly called "Silicon Valley." Until the mid-1960s, HP remained a specialized instrumentation company although their product range of instruments increasingly had computational capabilities as instrument controllers.

It is believed that HP had initiated discussions for taking over or investing in some computer companies of that time including DEC and Wang.[17] In 1964, HP's Dymec Division, which was tasked with making the computers incorporated into HP's instruments, was working on using DEC's computers. HP in 1964 bought out a software company from Detroit, called Data Systems Inc. By 1966, HP had introduced its own range of small computers, the HP 2000 series and shortly thereafter, in 1968 made its first desktop scientific computer, the HP 9100A.[18] Other developments followed rapidly and by 1972 the company had made its first business computer, the HP 3000. HP would go on to be a major computer and computer printers company. It also acquired other companies such as Apollo Computers, Convex Computer Corporation, and somewhat controversially, Compaq Computers in 2002. Dave Packard would go on to become the secretary of defense in the United States.

The floodgates for minicomputer manufacture were now open. Standardized and miniaturized electronic components from several companies in different countries were now available. Software to operate these computers was now being developed by several vendor companies. Many companies from around the world were now offering these minicomputers. There was RCA (later sold to Sperry Rand) with its "Spectra 70 systems" and Scientific Data Systems (acquired by Xerox) from the United States. The Japanese had Hitachi, NEC, Oki Data, and Fujitsu. Norway had Norsk Data. Germany had Nixdorf Computers (later to be acquired by Siemens). The Swedish aircraft and automobile conglomerate, SAAB, had its company Datasaab, Denmark had Regnecentralen, Italy's Olivetti was into minicomputers as also Electronics Corporation of India (ECIL). But the global markets were now demanding even smaller machines for use in small offices, and at home. A new era in computing was about to begin!

Personal Computers (PC)/Microcomputers

Until the beginning of the1970s, computers were still quite large in physical size and were predominantly meant for corporate or institutional use. What the market was demanding at that time was a smaller computer that would be a general purpose, stand-alone machine that could be operated by an individual with little or no formal computer training and needed to be at a cost level that could be afforded by a larger segment of the market, particularly individuals. Hence, the term applied, "personal computer."

No such "product" really existed as a commercially available one in the market. True in 1968, the Stanford Research Institute in California had demonstrated a small machine but this was only just better than a glorified word processor. By 1970, an IC manufacturing company also in California, Intel, about whom we will read later in chapter 9, had converted a set of ICs meant for electronic calculators into a single component that could function like a composite CPU, as defined by the von Neumann computer architecture. Intel termed this components as a "microprocessor," and the part number given to it was 4004.

By 1971, there was an updated version of the 4004, which was termed as the Intel 8008, microprocessor. The very first to spot the extraordinary potential of this new microprocessor was, believe it or not, not someone in Silicon Valley but a French Engineer, Francois Gernelle, working for a company in France called Intertechnique, which had received a contract to develop a small computer for the French National Institute for Agronomic Research. This institute could not afford a DEC minicomputer. Gernelle developed the Micral N for use by the institute. This computer became the world's first full microcomputer— not sold as a kit. Gernelle went on to establish his own company R2E in the year 1972.[19] The company R2E later became a part of the French computer company Bull.

Shortly thereafter, in 1973, the Palo Alto Research Center (PARC) of Xerox, as an experiment, designed their "Alto," microcomputer, which had for the first time a "graphic user interface," which would ultimately lead to the design of Microsoft's "Windows" software. The Alto became quite popular with research establishments and universities.[20] The Alto also introduced the computing world to the "mouse," an attachment that with cursor movements and clicks could control the operation of a computer.

At about the same time, the Scelbi Computer Consulting Company introduced a microcomputer kit, the 8H, at a price of around $550. Later in the 1970s, Scelbi figured that they could make more money selling books on computers than hardware kits and consequently got out of making kits and computers entirely.

By now, Intel had made an even better microprocessor, which had ten times the power of the 8008. The work on this started around 1972 and by 1974 was made commercially available. This new microprocessor enabled companies to make small computer systems by integrating standard bought out terminals, operating systems, and system software. This was a great boon to smaller companies who could now bring to market their own products.

Possibly, the first to use the new Intel 8080 microprocessor was another of those US high-tech companies that started in a garage. This one, Micro Instrumentation and Telemetry Systems (MITS), was started in 1969 in Albuquerque, New Mexico, by a former US Air Force engineer, Ed Roberts for making parts for hobby rockets. As part of a diversification program, the company began the manufacture of affordable electronic calculators and kits for their assembly.

When the IC supplier for their calculators decided to go into the business themselves for a range of calculators, MITS was forced to look for other product lines. By 1974, the company had developed a complete microcomputer in kit form using the new 8080 microprocessor and dubbed it the Altair 8800 computer. The electronics hobby magazine *Popular Electronics* featured this product on its cover in the January 1975 issue and the headline of the editorial read, "The Home Computer Is Here." MITS was now deluged with orders.[21]

Unfortunately, it appears that Ed Roberts was not as good a businessman as he was an engineer. He made a series of wrong marketing decisions including appointing dealers and distributors that could not deliver good results. To make matters worse, several competitors had also come up eating into a market developed by MITS, which now faced severe financial problems. Finally in 1976, MITS was sold off to the computer peripheral manufacturer, Pertec, which itself was taken over by Adler, a German company.[22] The one significant feature of the Altair computer was that its software was written in the now popular "language" called BASIC (Beginners All Purpose Symbolic Instruction) and was done by a certain Bill Gates from at that time a little-known company called Microsoft. We will read more about these two names a little later in this chapter.

One of the principal competitors of MITS and the Altair computer at that time was a company that started off as a computer consulting company called, IMS Associates, founded in 1973 in San Leandro, California, by William Millard. The company had the idea of putting some of the Altair-type computers into a more powerful computing system by use of networking the computers. Since Millard could not get hold of adequate numbers of the Altair computers, he decided that his company would go into the microcomputer business for themselves.[23]

The name of the company was now changed to IMSAI, and a microcomputer based on the Intel 8080 was designed and positioned as an improvement on the Altair. This machine was the IMSAI 8080. Unfortunately, IMSAI had not quite contended with the fact that good-quality software was required to get the machine to give optimum performance. They did not have Bill Gates and his company Microsoft with them.

With a very good distribution network supporting them, IMSAI did do well for a short period of time. But Millard, being an entrepreneur at heart decided to set up another entity, a sort of computer hobby store that initially was called Computer Shack and later the name was changed to Computerland. With Millard spending more of his time and resources on this new entity, the business of IMSAI began to suffer. The company, shortly thereafter, had to file for bankruptcy.

Sadly, as we have noted above, the pioneers in microcomputers, R2E (Micral), MITS (Altair), and IMSAI, whatever the merits of their technological developments, did not survive for long in the demanding marketplace. What were needed were companies with not only excellent hardware and software but also a sound overall ecosystem including good technology, good management and adequate financing. More importantly, what was required was a "vision" for the future in computing. One such company was to be established by, believe it or not, a survivor from the notorious Nazi "Auschwitz" concentration camp in wartime Poland.

Idek Tramielski, a Polish national, being of the Jewish faith, was incarcerated in Auschwitz along with his parents. His father did not survive the ordeals of the camp but he and his mother did and shortly after the war migrated to the United States. Now with a new name, Jack Tramiel, he enlisted in the US Army. On leaving the army, Tramiel established a company in Canada in 1955 called Commodore International, a good military sounding name, to make typewriters and calculators. By 1960, the company was doing well enough for it to go public. Very soon, however, Commodore's first venture capital financier got into serious trouble with the Canadian authorities.[24]

In 1966, Commodore had a new venture capital financier and also a new item in its product line, a pocket electronic calculator. However exactly as in the case of MITS, the chip supplier for their calculators, TI, decided to go into the calculator business for themselves. Thus in 1975, Commodore looked to be in dire trouble once again. Possibly as a result of his survival instincts honed at the Auschwitz concentration camp, Tramiel did not give in but again like Ed Roberts of MITS before him, decided to move aggressively into the microcomputer business.

In 1975, a company called MOS Technology based in Pennsylvania had designed a new microprocessor chip, the 6502, which was available at a cost considerably lower than what the market leader, Intel, was quoting. Not only did Commodore design a microcomputer around this, but also it ended up taking a controlling financial interest in MOS Technology so as not to be caught short as happened with their efforts with electronic calculators. In 1977, Commodore launched its PET microcomputer and over the next few years many other products were to follow including the Commodore 64—for a long time the best-selling microcomputer and the VIC 20, the first color microcomputer to sell for under $300.[25]

By 1982, things had started to unravel and get murkier at Commodore. Tramiel suddenly resigned from Commodore but returned shortly thereafter, only to leave again in 1984. A version of this chain of events seems to indicate that he was eased out of the company for buying out in private a controlling interest in an electronics, arcade, and video game producer, Atari even though officially Commodore was in discussion with Amiga, a financially stressed competitor to Atari. It is also believed that Tramiel lent money to Amiga in a bid to ensure that Commodore did not get control of Amiga.

Commodore did eventually manage to acquire Amiga, and successfully launched several new models over the next few years. But faced with increasing competition in the microcomputer field and with great volatility in the electronic games business (the arcade systems having been long forgotten), Commodore started facing myriad problems. It did not help that there were frequent management changes at the top of the company. To save costs, overseas factories in Philippines and Scotland were closed and many of the US-based plants were downsized. By April 1994, the company had gone bankrupt and a year later was acquired by a German company by the name of Escom. Tragically, Escom itself went into receivership in 1996. Tramiel's Atari also did not survive beyond 1996. It was sold off to JTS, a manufacturer of computer peripherals and finally in 1998 the intellectual property of the company was acquired by Hasbro.[26]

In 1976, a brilliant, young, 21-year-old engineer working at Atari decided to quit his job. Not only that, but also this young man called Steve Jobs persuaded a friend of his, Steve Wozniak, then working for HP, to resign. The two of them started a new company named Apple Computers, with "operations" out of Steve Jobs's garage in Cupertino, California. Another one of now America's famous high-tech garage start-ups had been born.

Wozniak used to attend meetings of the "Homebrew Computer Club," where he became fascinated with an Altair computer. Since he did not have the resources to buy such a machine, he decided to build one for himself. Having finished the design, Wozniak offered it to his then employers HP but was turned down. By April of 1976, Apple (as the new name of the company was) had made a machine running BASIC software, and called it Apple I.[27]

The Apple I computer was demonstrated at the Homebrew Computer Club in May 1976 and the first order for 50 of these machines was obtained. But the company was already on its way to making an improved version complete with color graphics, sound, as well as games, with the full machine housed in a nice looking single casing made of colored plastic. This was the Apple II, which really got the microcomputer market excited. However, the young team at Apple had precious little resources by way of finance and so offered this machine to Commodore, who turned down the offer as they were working on their own black-and-white display product, the Commodore 64.[28] The rest as they say is history! Sales of Apple II zoomed and the design for an even better microcomputer was in the works at Apple.

Apple now commenced work on two different computer series, the Lisa (an upgrade to the Apple II) and the Macintosh (an entirely new concept of a microcomputer). The Lisa came bundled with some exceptional software in addition to the operating system and was launched in 1983. But despite the publicity hype, due to the very high price the machine failed to bring in high enough sales. By 1986, the Lisa was discontinued.

The Macintosh project directly under the supervision of Steve Jobs came up with an extraordinary product. It was the first commercial computer to run a graphical user interface. It also had a sort of controlling device called the mouse. The Macintosh or the Mac as it was popularly known as, was a great looking machine with some good features but had drawbacks as far as usable office software applications were concerned. Further, for the Mac to be able to do excellent graphics and several other applications that only a few of the average PC users really wanted, Steve Jobs (described by an Indian friend of his, Vivek Ranadive, as the "da Vinci" of our times), with his penchant for perfection, went a little overboard on the design. As a result, the Mac became substantially more expensive than the products of other companies.

A big setback for Apple and the Mac was that their computers, although brilliantly engineered were not compatible with the easy to use software developed by Bill Gates and his team at Microsoft, which was available on other computers. This was called the Windows 3.0 operating system. There was also a management tussle brewing at Apple. In 1983, the board of directors had brought in a new CEO, a professional by the name of John Sculley, to sort out the growing problems at the company. Inevitably, this led to a face-off between Sculley and Jobs, and as a consequence in 1985 Steve Jobs resigned and went away to do another start up called NeXT, which would manufacture upmarket computer workstations targeted at the higher education market.

In a strange twist of fate, about a decade later, Apple now under a new CEO bought out NeXT as well as the operating system software that it had developed. Steve Jobs was brought back first as an advisor and in 1997 as the CEO of the company. It was also announced that the company would work with Microsoft's office application software. Starting in 2001, the company announced some revolutionary new consumer electronics products (about which we will read in chap. 7) and was now well and truly on its way to becoming one of the great electronics companies of our time.

Several other computer companies at one time or the other came up as potential challengers to the likes of Commodore and Apple. One was Timex Corporation a company specializing in watches and other time keeping products. They developed a joint venture with the British Company Sinclair Research, which had been started by the inveterate British inventor Sir Clive Sinclair who had developed and very successfully sold the ZX 81 microcomputer in Britain. Timex Sinclair introduced the TS 1000 computer in the United States in 1982 and at under $100 was then the cheapest home computer available. The company then introduced two other models, which were only partially successful. By 1984, Timex decided that they were no longer interested in the computer market, having had a somewhat bad experience in the bitter price war with Commodore. Timex from then on concentrated entirely on its horological products.

Radio Shack was a company started in Boston in 1921 and specialized in selling amateur radio communication equipment. The company went through difficult times especially with the downturn in the market of its best-selling line of citizen-band radios. They went bankrupt but were promptly bought out by Tandy Corporation, which had its origins in a thriving leather business. Tandy (Radio Shack) came up with their first computer the TRS 80, in 1977 shortly after Apple had launched its computers.

Tandy launched a few more products in the next few years and also acquired some companies to strengthen their product offering.

Unfortunately, the company could not withstand the rapidly growing competition and in the early 1990s sold off its computer related business to AST Research. Despite having a good product line and having actually ranked 431 on the Fortune 500 list of companies in 1992, AST floundered and was bought out by Samsung in 1996.

Adam Osborne, was born in Thailand in the year 1939, but spent most of his youth in the state of Tamil Nadu in India, where his father was involved in pursuing his interests in the Hindu religion. Osborne, with degrees from Birmingham University and the University of Delaware, became extremely interested in computers and more so in technical writing on the subject. Osborne became a prolific writer on computers and computing. In 1979, he sold his flourishing publishing business to McGraw Hill and with the proceeds set up Osborne Computer Corporation in January 1981.[29]

By April 1981, the company had introduced the Osborne 1, a computer with some exceptional specifications. It was portable, had a carrying case, and the size of the package was such that it would qualify to be placed under a standard airline seat. It also had its software fully paid for and included in the package (bundled). This product, acknowledged as the very first "laptop," was extremely successful to begin with and the company became a million-dollar entity in a very short time. But much like some other computer companies, Osborne ran into managerial, marketing, and software problems. By 1984, just three years after starting, the company had become bankrupt.[30]

Adam Osborne returned to India to set up a software company Noetics Software. His health, however, started to fail him and so Adam Osborne (sometimes described as "Vellaikara Tamizhan" or the "White Tamilian") went to live with his sister, a resident of the Indian hill station of Kodaikanal, where he died at home in 2003.

Then there were, or are, the other microcomputer companies. There was Compaq (started in 1982, acquired DEC, and now a part of HP), Dell (started in 1984 by Michael Dell as a direct sale vendor), Amstrad UK (first computer in 1984, acquired the "Sinclair" brand but now only producing TV set-top boxes), Gateway Inc. (started in 1986), and a few others from several different countries around the world including Hindustan Computers (HCL), Zenith, and PSI from India. But the whole PC and microcomputer business would come to be dominated by one company, the "Big Blue," IBM.

For a few years IBM had watched a whole new category of computers, the single-user microcomputer, become the rage in the market. It's not that they had not made attempts at bringing out a product specifically to address this particular market segment. In 1973, they had initiated a project for the development of just such a system. It was called the "Special Computer APL Special Portable," SCAMP for short.[31] After a successful demonstration, work on the IBM 5100 portable computer was started in 1975 with a view to providing computer capabilities to individual engineers and scientists. However, the weight (over 50 lbs.) and the cost ($9,000 for an entry-level system) were hugely negative factors and the machine was withdrawn from the market.[32]

An upgrade was now developed in the form of the IBM 5110, which was targeted at the corporate business market with general accounting capabilities. Even though this machine received large numbers of orders, when combined with the requisite peripherals it was still considered too big to qualify as a single-user machine and by 1982 this was also withdrawn from the market. Other models with revised specifications and capabilities were launched in quick succession but neither really caught the fancy of the market. At one time, it is believed IBM even seriously thought of buying out Atari. An analyst was reported to have remarked, "IBM bringing out a Personal Computer would be like teaching an elephant to tap dance."[33]

So was IBM up to the challenge? A very special project team called "Project Chess" was set up at IBM in Boca Raton, Florida, and the product to be delivered was given the code name of "Acorn." The team was tasked to completely turn the IBM philosophy of producing everything in house on its head. Discussions were also held with Bill Gates and his people at Microsoft. In a matter of a few weeks, by April 1981, the design of the machine was ready. Outside vendors were used for sourcing pretested parts and subassemblies, as there was no time to do this in house. Microsoft and other outside vendors were used for the operating system and application software. All vendors had to conform to rigid quality standards as the goal was "zero defects."[34]

Incredibly then, within a span of well under a year, on August 12, 1981, the IBM-PC (IBM 5150) was presented to the world. Powered by an Intel 8088 microprocessor, it had memory, disk drives, color graphics, and music capability, could hook up to the home television set, play games, and do office applications, as well as expansion slots all for the incredible price of $1,565! The distribution channels for this wonder of a machine included department stores like Sears-Roebuck, outlets like Computerland in addition to IBM's own product centers. The PC battle had been well and truly joined! A few months later *Time* magazine named the IBM-PC "Man of the Year" and splashed its picture on their cover!

The IBM-PC was successful beyond all imagination. One of those on the original design team, Mark Dean, on the thirtieth anniversary of the IBM-PC remarked, "Little did we expect to create an industry that ultimately peaked at 300 million unit sales per year." All competition had to bow before IBM. If, like Apple, your product was not IBM-PC compatible you did not stand a chance. Many manufacturers just gave up.

The problem now was that the IBM-PC had become just another commodity item. By December 2004, IBM decided to sell off its Personal Computing Division to a Chinese company, Lenovo, allegedly with links to their Peoples Liberation Army for a total transaction worth $1.75 billion. Just as well, for computers were getting even smaller, they were going the way of the Osborne 1. The era of the laptop had arrived.

Although Adam Osborne is generally credited with being the first to make a laptop computer, there were others also working toward that goal. In 1981, Epson a Japanese printer company introduced the HX 20, a battery-powered portable computer. In 1983, Radio Shack brought to the market the TRS-80, Model 100 machine, which was based on a design done by Bill Gates and Kazuhiko Nishi of Microsoft and three years later, the Model 200.

IBM itself came up with its first laptop, the IBM 5155 in 1984 and Compaq made its first in 1988. It is, however, to the credit of the Japanese company, NEC, to have come up in 1989 with the world's very first "notebook"-style portable computer, the NEC "Ultralite." Other laptop products followed from Compaq, Zenith, and Apple, and the Japanese company Toshiba would be the first to mass-produce laptop computers. Shortly after IBM sold off its PC business to Lenovo, the once industry leader, NEC also exited the PC business by forming a joint venture with Lenovo.

Computers were now downsizing even more to a size of what were earlier called as "personal digital assistants" (PDAs) and subsequently as "tablets." We are now into contemporary times as these "handheld" items rapidly became a part of consumer electronics and out of mainstream computers (see chap. 7). Computing itself is now rapidly moving toward what is called "cloud computing" where most software applications, data, and information are held in large server banks (the "cloud").

Software

As is very obvious, a computer by itself can do precious little. With just the electronics inside a box any amount of inputs from an entry device such as a keyboard will at best give some meaningless stuff on a display system exemplifying what most of us have been taught: "Garbage In—Garbage Out!"

A computer needs some instructions or commands about what it is supposed to do and how it is supposed to operate. This is done by what is termed "software." In the mechanical calculators and tabulators of olden times, the instructions of what the machine should do were pretty much built into the mechanical keys and linkages inside the machine. One pressed a key or tab and the machine was forced to do an operation that was already "embedded."

In its most basic form, software is nothing but an organized, structured system of logical statements. In ancient India, which gave the world the decimal system, algebra, and trigonometry, these logical statements were used by philosophers and mathematicians to solve complex mathematical problems. These go back to Panini in 500 BC or so and to Pingala in 300 BC. Subsequently, in what may be described as the "Classical Period," AD 400 to 1200, Aryabhata and Bhaskara, two outstanding mathematicians of their time, were writing complex algorithms (much like for present-day complex calculations and for search engines).[35]

Modern-day computer software may, however, deemed to have begun in 1945 with the "Plankalkul" of Konrad Zuse, as described earlier in this chapter. In 1948, an engineer and mathematician from MIT, Claude Shannon, who had earlier done pioneering work on Boolean Algebra applications in electronics, wrote his now famous treatise on "A Mathematical Theory of Communication," while he was working at Bell Telephone Laboratories. It is this paper that laid the foundations of modern information theory and the basis of writing computer software or "coding."

Now, of course, computers cannot understand our normal language. They can only understand the binary language of zeros and ones. A combination of these may be made for a computer to understand, for example, "add" operation. But writing any statements with just zeros and ones is hugely cumbersome and impractical and hence, the requirement of intermediaries such as "compilers" and "assemblers." These, simply stated, are able to convert plain language into machine language.

The credit for possibly the greatest impact on computer software in modern times must go, strangely so because of the time period, to a lady, Grace Murray Hopper, a graduate in mathematics and physics from Vassar and subsequently a PhD from Yale University in 1934. In 1943, she joined the US Navy as an officer and was assigned to work with Howard Aiken at MIT on the Mark I digital computer project, which had applications in Naval Ordnance Development.

While at Harvard, Grace Hopper wrote the first compiler (an intermediary statement that enables instructions written in normal language into computer code). Hopper, followed this up with writing the "Flow Matic" program, which is described as the very first English language data processing compiler and was later used to program the UNIVAC I and II computers.

Grace Murray then went to work with the EMCC, which as we have noted above got acquired by Remington Rand and finally merged into Sperry Corporation. Grace Murray continued in the Naval Reserve and attained the rank of admiral before finally retiring from active work.

To make life easier for the computer programmer, it was felt that coding may be done in what are called higher-level languages where normal sounding words and syntax may be usable. The first of these was FORTRAN written by John Backus of IBM and made commercially available in 1957. Subsequently, other languages came up such as COBOL, C, Pascal, and Prolog. But perhaps the one to have the most impact was BASIC.

BASIC was invented in 1963 at Dartmouth College. However, it was only in 1975 when it came to prominence after Bill Gates and Paul Allen at Microsoft wrote programs for Altair in a version of this language, and later for IBM and Apple as well. After a few years, BASIC's popularity waned but came back in 1991 with the development of a version called "Visual Basic" (VB) by Microsoft. Subsequently, many different and updated versions of VB would follow.

The most important program for a computer system is what is termed as the "operating system." This is the piece of software that "directs" other programs as also the various subsystems such as the display, keyboard, memory, and peripherals about what to do by sending appropriate "commands."

In 1969, Bell Laboratories developed the UNIX operating system and is still operational. In 1979, a company called Intergalactic Digital Research introduced the CP/M program, a very simple microcomputer operating system. But the really most important developments as far as operating systems go came with the setting up of the now software giant, Microsoft.

Childhood friends, William (Bill) Gates and Paul Allen, spent a lot of their time in the school (Lakeside Prep School) computer room from 1968 onward. They even gave some normal classes a miss in order to spend time with computers. The school had a Teletype terminal that was connected to, reportedly, a GE computer at a remote location. It is said that the duo hacked into the computer but instead of expulsion, they were tasked to improve the school's computational facilities. The company that owned the computers gave them free computer time in exchange for technical tips.

The pair also set up a company, Traf-O-Data, in their hometown of Seattle, Washington, and sold a computer to the city authorities for counting traffic. The duo was also hired by a local company called Information Sciences Inc. to put together a payroll program for them.

Paul Allen, the elder of the two, appeared for the Scholastic Aptitude Test and got the maximum possible score. He then joined Washington State University, but promptly dropped out to become a programmer with Honeywell. Bill Gates, the junior by two years, went on to join Harvard University in 1973 for pre-law studies. In 1974, Allen saw a photograph of an Altair 8800 microcomputer on the cover of the magazine *Popular Electronics* and was convinced that the microcomputer would be the business of the future and software would be in great demand. He persuaded Gates to also drop out of university and to go into business together.

Bill Gates phoned the computer company MITS and offered to write a special version of Dartmouth College's BASIC programming language. Incredibly, MITS agreed. Within eight weeks, the two friends had the product ready and demonstrated it to MITS. The company was so impressed that they offered to sell the product as Altair BASIC.[36] With this success behind them, Gates and Allen set up their company Microsoft in November 1976. The next couple of years were a bit rough for the new start-up although they had by now a good team of developers.

The breakthrough came actually as a result of the endeavors of another Seattle-based company. In April 1980, Seattle Computer Products decided that they would develop a new operating system for its microcomputers as the expected CP/M operating system was overdue. By August of 1980, they had developed the Q-DOS (Quick and Dirty Operating System), a sort of hurried job, but which worked. Later that same year, an updated version was developed and Microsoft took up nonexclusive rights to this operating system.

By July 1981, Microsoft bought out the full rights to this operating system, tweaked it somewhat, and called it MS-DOS. Shortly afterward, IBM announced its PC and the operating system that came with it was MS-DOS from Microsoft. The company went public in 1986 making Gates and Allen billionaires. The very next year the company introduced the pathbreaking Windows system and by 1993, a million copies of this were being sold every month.[37]

Now, of course, these operating systems from UNIX to Windows came at a cost as a young Finnish student in Helsinki realized in 1991. Linus Torvalds had bought himself an IBM-PC when at Helsinki University, but wanted to use UNIX instead of the standard MS-DOS operating system. He could not afford the huge asking price. Even the abridged edition of MINIX was far too expensive. So this young man decided that he would write his own. Thus was born the Linux operating system. Torvalds being somewhat of a "do gooder" decided that his development should be freely available to all. Linux was made available under a General Public License and arguably set in motion what is today called the "open source movement."

Of course, the phenomenal growth of computers and computing brought along with it some not so pleasant side effects. With the rapid spread of software expertise, there were many who started to use their technological prowess. Some for mischief, some for criminal activities, and others, including state-sponsored actors, for the age-old art of spying and espionage. Thus was born a whole new cyber science of computer viruses, malware, "worms," and what have you. A fascinating subject for another book maybe!

CHAPTER 7

Media Recorders/Players, Mobile Phones, Smart Devices, and Tablets

If music be the food of love, play on.
—William Shakespeare, *Twelfth Night*

Good Moses had tablets before Apple and Samsung.
—Anon

What good is it, besides surfing the web in the bathroom?
—Steve Jobs, on the Apple tablet prototype

Man has for long wanted to have music readily available when there was a desire to listen to it. In the old days, the rich and powerful could always order or hire someone to do this in person. As for the others, well, either you attended a concert, did your own thing, or maybe, whistled to yourself. This, however, changed in 1877 when Thomas Alva Edison invented the cylinder phonograph.

Edison was working on his development of the telegraph and had the need to record messages, which he did through indentations on a kind of paper tape so that they could be repeated or be referenced subsequently. He then wondered if something similar could be used to record telephone messages. He came up with a contraption comprising a metal cylinder with tin foil wrapped around. He then incorporated two sets of a diaphragm with an "embossing" point, one for receiving and the other for sending. Sound vibrations coming through a microphone would then indent the foil on the respective cylinder. Edison tested his machine by speaking "Mary had a little lamb," and found he could play it back exactly. The world now had its first phonograph!

In 1878, a cornet player by the name of Jules Levy had made the first full recording for a phonograph cylinder. He played "Yankee Doodle." In 1879, Alexander Graham Bell improved on Edison's invention by using a wax record, which was cut by a chisel-type stylus. He called this the "Graphophone." The rights to this invention were first acquired by the Volta Graphophone Company, then successively by the American Graphophone Company, North American Phonograph Company, and Columbia Phonograph Company. The last of these would become Columbia Records.

In 1887, an American inventor, Emile Berliner, came up with his invention in which he managed to etch fine grooves into a flat-disc zinc cylinder on the input of sound signals. He called his invention the Gramophone and received a patent for it. Berliner started The Gramophone Company, to manufacture the machine as well as the discs. He used an image from a painting of a little dog (named Nipper) listening to the sound device as the official trade mark of The Gramophone Company. He called this "His Master's Voice," and the image later became an international hit. Berliner sold the licensing rights to his machine and the method of making records to Victor Talking Machine Company, a company set up in 1901 by Eldridge R. Johnson. Victor Talking Machine Company was subsequently acquired by RCA in 1929 and the name of the company changed to RCA Victor—a name familiar to many who bought phonograph records in the 1960s.

Record playing machines would over the years undergo major improvements as electronics technology developed. There were, of course, the standardized record speeds of 78–45–33 1/3 revolutions per minute (rpm) and the availability of prerecorded music on these records. Amplification, stereo as well as noise reduction technologies followed over the years. Several companies got into manufacturing these players, but it was many Japanese companies including JVC, Kenwood, Sony, Pioneer, Sansui, and others that emerged as the big players in the global market.

The year after, in 1888, Edison upgraded his earlier design of the phonograph to incorporate an electric motor to drive the drum, but this modification made the machine quite expensive for that time. It was, however, a Danish inventor that would come up with a completely new approach to making recordings. Valdemar Poulsen, a drop out from medical college, took up a position with the Copenhagen Telephone Company, to pursue work on his interest of technology. He was aware that in 1878 an American inventor, Oberlin Smith of Ferracute Machine Company, in New Jersey, had published a technical paper[1] describing the possibility of recording telephone signals by magnetizing a steel wire. There is, however, no documentary evidence of Oberlin Smith ever having made an actual working recorder.

In 1898, Poulsen developed a system of recording sound by magnetizing a steel wire with the use of a small brass cylinder with a thin wire wrapped around it. The wire was energized by a battery when an output was received from a microphone. The device was called the "Telegraphone." This was the birth of magnetic recording technology.[2]

In 1903, Poulsen started the American Telegraphone Company in Washington, DC, to manufacture his product. The company sold several of these machines but predominantly as telephone answering machines or as Dictaphones. The company, however, closed down quite soon. The rights to the invention were passed to an English inventor, Louis Blattner, who modified it to try and synchronize sound with motion pictures. Having failed in his effort, he then sold the rights to the Marconi Company in Britain.[3]

The Tape Recorder

The breakthrough in audio sound recording and playback came in Germany in 1927. An engineer, Fritz Pfleumer figured that he could put magnetic stripes of iron oxide onto thin paper using lacquer glue, which would be a replacement for the cumbersome wire earlier used for recording. Pfleumer subsequently received a patent for this invention. In 1930, the German company AEG started work on developing a recording machine using Pfleumer's work. In 1932, AEG received the rights to Pfleumer's invention and at the same time worked with the company BASF to actually make some usable magnetic tape.

In 1935, during Berlin's "Radio Fair" exposition the world's very first tape recorder called the "Magnetophone" made by AEG was demonstrated. It is said that the world did not really know about the technological advancement of the Magnetophone until after the German surrender in WWII when in 1945 a few of these machines were found by American troops and shipped to the United States for close scrutiny.

By 1947, a company in New Jersey, Rangertone Inc. had made a professional tape recorder based on the Magnetophone design. AEG itself merged with Telefunken in 1967 by which time it had manufactured almost two million tape recorders. AEG Telefunken finally ended up being bought out by Daimler in 1985. In 2005, Electrolux of Sweden acquired the rights to the AEG brand.

Although we now had two practical methods of sound recording and playback, the machines to actually make this happen were not really something you could carry around. They were "clunky" and definitely not portable, yet for several years they gave so much joy to so many people. In 1939, a Chicago engineering student, Marvin Camras, had developed an improved and inexpensive "Telegraphophone." In 1940, he joined the research foundation of the Armour Institute of Technology (later to be known as the Illinois Institute of Technology), in Chicago where he worked on improved sound recording systems. During WWII, the research foundation received a grant from the US Navy to develop a more rugged and miniaturized sound recorder.

After the war, a group of engineers from the Armour Research Foundation left to start their own company, called the Magnecord Corporation and obtained a license from the foundation for technology developed there. In 1948, this company produced the PT-6, the first portable tape recorder. In 1949, Magnecord went on to make the first stereo tape recorder. Several other companies rapidly followed into the business of manufacturing tape recorders. These included Ampro, Pentron, Pierce, Ferrograph (UK), Roberts, Amplifier Corporation of America, and others. Ampex Corporation also obtained a license from the Armour Research Foundation and in 1948 produced their first professional tape recorder, the Model 200. With all this competition, Magnecord found itself in great financial difficulty. In 1957, it was purchased by Midwestern Instruments.

Another license of tape recorder technology from the Armour Research Foundation was given to a Chicago-based company, Webster-Chicago Corporation. This company was already manufacturing radios, battery eliminators, and other electronic equipment when in 1945 on acquiring the license, made its first wire recorders for the US Navy. By 1952, when it changed its name to Webcor, it had become the leading manufacturer of wire recorders in the United States.

The same year Webcor launched its first tape recorder meant for the consumer market, the Model 210. This was the first machine to have a pair of record/play magnetic heads as well as a pair of induction motors. This allowed the machine to play tapes in both directions without having to turn the tape reels manually over. Webcor became a leading name in the tape recorder business. They continued to make them as well as other consumer electronic equipment well into the 1960s when they ran into severe competition particularly from manufacturers overseas, which seriously impacted the fortunes of the company. In 1967, Webcor was acquired by a company called US Industries Inc. and subsequently in a complete reorganization in 1971 the name was changed to Webcor Electronics.

Outside of Germany and the United States, it was Japan that was now beginning to catch up in the technology. Tokyo Tsushin Kogyo, later to be called Sony, started to manufacture magnetite-coated paper recording tapes in 1950. By 1954, the company had manufactured its first tape recorder, the Type "G." Many others such as Akai, Sansui, and TEAC, would shortly follow, and would end up dominating this business.

Tape Recorders of that age were described as "reel to reel" as a large spool (reel) of magnetic tape loaded on one reel would unwind through a magnetic head (which would pick up the recorded content for playback) onto a receiving reel. The machines were quite large and definitely not what today would be called portable. Further, handling the spools of tape was a cumbersome process and storing stacks of them in their covers took up too much space. A technological change was now needed.

Cartridge/Cassette Recorders

In 1958, RCA brought to the market a reel-to-reel cartridge that had taken them a few years to develop. It was still quite large in dimensions and failed to make an impression in the market. The following year, in 1959, Collins Radio introduced the "Fidelipac" cartridge, an ungainly looking product aimed principally at the broadcast market and not at the general consumer.

The most significant change came in the form of a development made by Philips. In 1962, Philips invented a tape system in a small, closed plastic housing. The tape itself was only 0.15 inches wide and would wind /unwind over two small plastic rollers placed inside the housing. The full tape package was just 4 inches in length and 2.5 inches in width. Philips called this the "compact cassette." It would drastically change the tape recording and playback business!

This development brought about a profound change in the whole eco-system of audio tape recorders. The machine, also designed by Philips, was not much bigger than a hardbound reference book. The tape could easily be flipped over and play content recorded on both sides. Most importantly, it could run on standard torchlight batteries and hence could be toted around. The use of high-quality BASF magnetic tapes and some smart electronics designing ensured a reasonable quality of performance.[4] The portability of the device and the ease of handling the small cassettes enabled people to carry and play their music at their convenience.

Taken aback by the success of this revolutionary new product, Grundig threatened to work with Sony in developing a competing system. Fortunately, good sense prevailed all around. Philips agreed to waive off royalties on their invention and proposed a sort of open general license. Large numbers of manufacturers came on to the bandwagon, and as a consequence the Philips's compact cassette enjoyed an incredibly successful run all the way up the mid-1980s.

The portability of the cassette-based tape recorders/players enabled some interesting variants to be made. The first was the development of what a certain section of American urban society popularly called the "boom box." This double speaker "souped" up tape player was developed by Philips in 1969 and rapidly became a cult item. Other manufacturers including Marantz, Sony, and GE soon followed.

Sony, however, felt that the standard cassette player was still too large for certain applications such as listening to music through a headphone on an airplane or while out walking or jogging. In 1978, Sony came out with a small portable cassette tape recorder, the TC-D5, which produced great stereo sound. The then chairman of Sony felt this product was still too bulky and the size of the headphones was inconvenient. The design team at Sony then reengineered this product, by taking out the recording function and substantially reducing the headphone size to become an earphone. On June 21, 1979, Sony introduced this portable cassette player to the market. It was called the "Walkman" and almost overnight became a sensation. In under a decade, about 50 million Walkmans were made and sold.

But Sony was in for a bit of a surprise. Shortly after the launch of the Walkman, they were hit by a lawsuit filed by a Brazilian-German inventor by the name of Andreas Pavel. Pavel, in his lawsuit claimed that he had already invented such a device as early as 1972 and had even offered it for manufacture to Yamaha and Philips among others. Pavel's invention dubbed the "Stereobelt" had patent applications filed in Germany, Italy, and even Japan. Sony settled with Pavel out of court and it is believed a payout of $10 million was made.[5]

With the advent of compact discs (CDs; see below) and other digital media including direct music downloads onto computer memories, the cassette tape recordings, much like the long-playing records of yesteryear went into technological oblivion with the exception of a few retro productions recently made as collector's items.

The Video Recorder

The early tape machines recorded and played back only audio content. There was a growing need for recording and playback of a combination of audio as well as visual or video content. The great crooner, Bing Crosby, had a company called Bing Crosby Enterprises (BCE). Bing Crosby was very keen to have good tape recorders for his very popular television shows, as the disc recorders they were using were not very good. BCE even invested money in Ampex and in return became the selling agents for Ampex in the western part of the United States. The financial input from BCE helped Ampex, a company established in 1944 by Alexander Matthew Poniatoff, make their first machine, the Model 200, in 1948.[6]

The BCE Electronics Division was given the task of developing a good videotape recorder system for use in the movie, television, and entertainment business. Using a modified Ampex Model 200 tape recorder, they demonstrated in 1951 possibly the first public recording of a video signal but the results were far from satisfactory. An improved version was brought out in 1955 but was still not good enough. By then, Ampex had made a better product.

BCE bought one of the Ampex machines and stopped work on their own version meant specifically for the movie entertainment business. Their product, however, was very successfully adapted for the growing instrumentation and defense business. BCE had over the years, worked with 3M on the development of high-quality magnetic tapes. 3M ultimately ended up buying out the BCE Electronics Division.[7]

It was Ampex that developed in 1956 the first prototype of a combo audio-video tape recorder, the VRX-1000. This would be the precursor to what would some years later be the ubiquitous videocassette recorder (VCR). However, this product was so large and expensive costing approximately $50,000 that except for some television studios there were no buyers. With the market for such machines being seemingly so large and attractive, other companies started to put in efforts in developing a machine that would attract the home user. Many companies did start development programs but it was Philips that introduced in 1964 the very first domestic video recorder.

The Japanese company Sony introduced their videotape recorder for home use in 1965. This CV-2000 videotape recorder used a 0.5-inch wide tape and was capable of over one hour of continuous recording and playback.[8] Shortly thereafter, in 1967, Sony introduced the very first portable videotape recorder, the DV-2400. It followed that up by developing a recording system using 0.75-inch tapes in the form of a cassette (U-Matic System) and based on this the world's first VCR, the VO-1600.

At about this time, a US company was developing an interesting product. Cartridge Television Inc., based in Palo Alto, brought to the market in 1972 the first American-made VCR that could play back prerecorded videotapes. These tapes had movie and other recordings obtained from some Hollywood studios. Their blank tapes could also directly record off a television set to which it could be connected. AVCO, a major defense contractor was roped in as a financial partner and a deal made with Columbia Pictures for movie content.

Large sums of money were spent in setting up manufacturing facilities for the cassette recorder/player, the magnetic head as well as for the cassettes themselves. A major advertising campaign was launched coast to coast and Sears Roebuck came on board as a retailing partner. Everything was just right, but something terrible happened. In November 1972, it was found that all the tapes started to decompose rapidly due to their poor quality. This had the potential of damaging the magnetic heads of the machines. Sales now had to be stopped completely.

It took Cartridge Television Inc. several months and a huge expenditure on making improved cassettes as also on another massive advertising and promotions campaign to make a comeback. With revenues still not commensurate with the huge expenses, Wall Street traders took a rather dim view of the prospects of the company. In 1973, there was a run on its shares, and AVCO decided to pull out of the project writing down some $50 million. Cartridge Television was now doomed and declared bankruptcy.

Thus, at a time in the 1970s, it seemed that Sony and Philips would become the dominant players in the rapidly growing VCR business. Philips had released the first VCR with a built-in television tuner in 1972 as also a color VCR in 1973. But Sony had another product under development. In 1975, they introduced to the market a better machine using a system they called "Betamax." That is when Sony ran into their first problem.

The Betamax system could not record any more than an hour of content off a standard television set in the United States, which used the NTSC system for its television broadcasting. The recording time for the European PAL system was, however, as much as three hours. The US market needed recording times at least as long as that of the aired broadcast of an American Football game, which the Betamax could not give.

Other manufacturers were now making efforts to develop their own systems. There was the "Video 2000" promoted by the team from Philips and Grundig. Sanyo had its "V-Cord" system and Quasar its "Great Time Machine." None of these could make much headway and were in danger of losing out to the only viable system at that time, the Sony Betamax.

Unknown to Sony, in 1975, another Japanese company, JVC (part of the Matshusita group), had also started work on their own VCR system.

JVC was set up in Yokohama in 1927, as the Japanese subsidiary of the US company, Victor Talking Machine Company, primarily for making phonographs called "Victorola," and subsequently, stereo record players. JVC could also foresee a bright future for a high-quality VCR that could deliver results particularly in the US market.

Sony and JVC had held some discussions about a possible license deal for the Betamax system. JVC turned down the offer from Sony and decided to bring their own system, the "Video Home System" (VHS), to market. In addition to VHS technology providing longer recording times, it also resulted in lower costs. As a result, all the other major manufacturers such as National Panasonic, Zenith, and RCA to name a few, jumped onto the JVC VHS system. Betamax was now doomed and went into oblivion and the VHS video recording format became the predominant system of choice.

Optical Discs/Compact Discs

By the mid-1980s, users of tape devices were beginning to tire of frequent snarls and snags in their tapes, whether audio or video. Despite efficient and well-designed machines, tapes would get jammed, twisted, or at times, just snap. Something more reliable was needed.

In 1958, Dr. David Paul Gregg, then a scientist at a California-based company called Westrex, had started work on an optical recording device. By 1961, he had filed for a patent for what he called the "Videodisk" in which he used electron beams to record content onto a coated disc. Gregg then left to work for the Mincom Division of 3M who for some reason felt that they would be better off delegating the work on videodisc technology to the Stanford Research Institute. Gregg and two of his colleagues quit Mincom to start their own company, Gauss Electrophysics. The company and its patents were acquired by the Music Corporation of America (MCA) in 1968. By 1972, MCA Discovision had launched its videodisc system and demonstrated the first copied videodisc.

First MCA Discovision worked with Philips to develop a laser disc system. This unfortunately turned out to be a failure. Then in 1977, MCA Discovision joined hands with Pioneer Electronics to form a new company by the name of Universal Pioneer Corporation (UPC) and started producing an industrial optical disc player. In 1979, IBM acquired a 50 percent shareholding in MCA Discovision and named the new entity Discovision Associates. Discovision would, a decade later, become a full part of Pioneer. Although all this great technological development did not result in any really successful products, the patents received by Discovision were most useful and formed the basis for the development of the Pioneer's laser disc, Philips's CD, and Sony's minidisc.

James Russell, a scientist and an avid music buff working at the Pacific Northwest Laboratory of the Battelle Memorial Institute, figured that the most efficient and optimum music system would be one that could record and playback without any physical contact between the component parts. This would necessarily require an optical system and using a digital system much like that in computers. Russell started work on such a system in 1968 and by 1970 Russell had developed and patented the first digital to optical recording system. More work on this system went on until 1985 by when Russell had received 25 patents for what he described as "Compact Disc–Read Only Memory" (CD-ROM) technology. This development came to have a profound impact on not only the computer industry but also on the music industry at large.

In 1969, a Dutch scientist, Klass Compaan, working with Philips came out with his concept of what he described as a CD. In 1970, Compaan with the help of a colleague, Pete Kramer, made a glass-coated disc from which information could be picked up or stored with the use of a fine beam laser. By 1972, the two had made a color version of this disc. Further development was initiated on this technology at Philips. On March 8, 1979, Philips introduced the very first CD audio player. It was on January 1, 1981, that BBC publicly demonstrated a prerecorded CD of the popular singing group, the Bee Gees.

Sony was also working on developing technology for CDs. In September 1976, Sony demonstrated its optical digital disc. Philips and Sony in 1979 decided to cooperate in developing this technology and products associated with it. In 1982, Sony brought to the market its CD player, the CDP-101, and Philips introduced their CD-100 shortly thereafter. In June 1984, Sony came out with the "Discman" the CD equivalent of the Walkman. Long-playing records (until their recent revival as a sort of retro item) and audio tapes were now history.

The CDs were great for audio content. However, for video content something different was required that would be able to take a much larger amount of digitally encoded data as video. In 1993, as a result of cooperation between JVC, Sony, and Philips a new format of CD, the Video CD (VCD), was introduced. It could record at a pinch about 80 minutes of a movie. The quality was poor but due to the very low prices it had a reasonable response largely as a video pirate's dream. Unfortunately, given the length of most movies one usually needed at least two of these discs for a full-length movie. For the more discerning market, the VCD was then not the answer and hence failed in the market place.

The following year Philips and Sony announced that they would collaborate on a new and better format for videodiscs. At about the same time, another group comprising of Toshiba, Hitachi, JVC, and others were working on a competing format. Fortunately, all companies saw the pitfalls in repeating the saga of the expensive and near ruinous battle of the videotape format wars between VHS and Betamax. They agreed to work together, with a little prodding from the likes of big companies like IBM, as the product would also have profound implications in the requirement of data storage for the computer industry. The result of this cooperation was the announcement in 1995 of what is called the digital video disc or also the digital versatile disc (DVD).The first DVDs were officially released in the market in 1997 to popular acclaim, and until the advent of the Internet (see chap. 8) was the way most people would get their recorded music.

The MP3 Format and "Gizmos"

Tapes, cartridges, cassettes, and discs all had the one fundamental weakness. They all needed moving parts in one form or the other. In many personal mobility applications such as listening to music while walking, more so jogging, this presented limitations on the lifetime of the playing equipment. Furthermore, the media involved, whether it be tape or discs, also had limitations. Tapes snagged and broke. Discs got scratched, collected dust and humidity and content got ruined. A more robust technology, within the requirements of size, cost, and portability was needed.

The solution came from a German institute, the Fraunhofer-Institut fur Integrierte Schaltungen (the Fraunhofer Institute for Integrated Circuits), based in Erlangen, Germany. This institute started research work on digital audio broadcasting in1987 as a result of an EU-sponsored project under the leadership of Karlheinz Brandenburg. Brandenburg, a specialist in electronics and mathematics, postulated that one could come up with an algorithm that would enable the compression of original sound signals by a factor of as much as 12 times without there being any loss of quality. With such compression coding, one did not need the traditional recording media and smaller and a more rugged solid-state replacement (called "flash memory") could be used.[9]

In 1988, a subcommittee of the International Standards Organization called the "Moving Picture Experts Group" (MPEG) was set up to standardize new formats for media use. In 1993, Fraunhofer's coding algorithms were incorporated as the MPEG I standard. Successive versions of this standard followed, and in 1995 the MPEG Audio Layer III (MP3 for short) was adapted. In 1996, a US patent for this technology was received.[10]

The Fraunhofer Institute tried to make its own media player in the early 1990s but sadly it was not successful. However, in 1997, a Croatian software professional by the name of Tomislav Uzelac created a decoder version of MP3 called "AMP," which was more successful. Uzelac set up a company called Advanced Multimedia Products (AMP) to commercialize his development. This company subsequently merged into Play Media Systems in which form it still exists as a California-based company.

By 1998, a little-known South Korean company called Saehan came out with the very first mass-produced flash-memory-based MP3 player, the "MPMan" (sold in the United States as Eiger Labs MPMan), a device not much larger than a cigarette packet. Shortly thereafter, in the same year, a US company based in California, Diamond Multimedia, brought out its "Rio" MP3 player. By now, the Recording Industries Association of America (RIAA) fearing wholesale loss of royalties on music content sued Diamond Multimedia but the court in its wisdom turned down the suit. In 1999, Diamond Multimedia was merged into S3 Incorporated but was never really very successful. By 2003, all its assets were purchased by Best Data. The pioneer of the business, Saehan, also closed operations in a few years and so did their partners in the United States, Eiger Labs.

The ignominious start to the introduction of MP3 players would, however, soon come to an end. A research team at Compaq's research center at Palo Alto, California, realized that just having a flash memory, as the common MP3 player of the time had, was a major problem. The very limited capacity made it not much better than a toy. What the gizmo needed was the equivalent of a computer hard disk drive but, of course, within the size limitations needed for a truly portable device.

In 1998, Compaq developed a device with a memory running into gigabytes as compared to a few megabytes in the earlier MP3 products. The following year this technology was licensed to the South Korean company, Hango Electronics, who produced and launched the "Personal Juke Box," the very first MP3 player with a hard disk. Unfortunately, Hango did not really do very much with this technology and possibly wasted a golden opportunity. Today, the company exists as Remote Solution Company Ltd., and remains a medium-scale entity making a range of remote control devices.

Compaq by itself was by then getting into some difficulties in the highly competitive small computer business. By 2002, it had been acquired by HP and the idea of commercializing of what could potentially have been a blockbuster item for the company was now more or less forgotten.

But along came this great genius called Steve Jobs! In 2001, the small computer business at Apple was sort of scraping along trying to find a market niche different from the ubiquitous IBM PC and its compatibles. In January of 2001, the company had launched "iTunes," an innovative program that could convert audio CDs into compressed digital format, and organize a music collection on the Apple Macintosh computer. With his fabled intuition and instinct, Steve Jobs felt that there was more to this music business than just providing iTunes for a computer.

Within the year, in 2001, Apple announced the launch of what it called the "iPod." A hard-disk-based MP3 device with a storage capacity of over one thousand songs with a size that could fit into a shirt pocket at an unbelievable price of just $399. In just two months, the company had sold 125,000 of these devices. In the following years, Apple introduced several newer versions with different specifications and price points—the iPod Mini (2004), the iPod Shuffle (2005), the iPod Nano (late 2005), and the iPod Touch in 2007. By now, Apple was on a roll and had captured not only the market (with some 90% market share) but also the hearts and minds of customers worldwide. Just look at the numbers! A total of over 350 million iPods sold, 16 billion iTunes downloaded, the price per Apple share zooming from $8.57 in June 2003 up to $700 in 2012! Surely, HP must clearly be ruing the great opportunity that they had in hand when they acquired Compaq but let go!

Mobile and Cellular Telephony

The world now had a solution for the requirements of music on the go. But it still needed a satisfactory solution for making telephone calls on the go—something that had tickled everyone's fancy since the time Chester Gould's famous comic character, plainclothes police detective, Dick Tracy, got his wrist radiophone in 1946. Over the years, there had been several attempts to make actual working models of the Dick Tracy wrist phone.

In 1942, a US industrialist, Diet Smith tried to get one made. In 1954, Sylvania came up with a prototype using transistors. In 1963, Davenport & Waldon, a Los Angeles company was actually advertising one for sale at $7 a piece. One such advertisement was carried in the June 8, 1963, edition of *Billboard* magazine, on a day that Pat Boone had the best-selling single record.

A recent flurry of media reports speak of a lady using a mobile phone look-alike gadget in 1938 that she claims was supplied to her as an experimental radio device by Dupont of Leominster, Massachusetts.[11] Though how that device would have worked without the requisite infrastructure is not explained. Looking at the date of the reports, it was possibly only an "April Fool" item.

There were, of course, the fans of the series of "communicators" from the *Star Trek* television series, the "Beam me up Scotty" ones. Clearly, a prop maker's imagination as the technology for telephony in space did not exist. Yet, the concept seems to have inspired several scientists in days to come.

Some types of "on the go" telephone communications were, of course, possible using radio sets. Personal radio communications started off in the United States in 1945 when permission was given to utilize a designated frequency band ("Citizen Band"). The popularity of Citizens Band (CB) radio communications grew substantially in the 1970s with long-distance truckers taking a leading part. Thanks to a few Hollywood movies featuring such radio sets, CB rapidly became a sort of cult item. But this was not the same as being able to make low-cost telephone calls at will and on the go as it were.

In 1946, a pair of Soviet electronics engineers, G. Shapiro and I. Zaharchenko purportedly tested a version of a radio mobile phone in their car that could connect to the local telephone network from up to a range of 20 kilometers.

In 1946, AT&T introduced the first experimental mobile telephony system. This used a single antenna and served a very small local geographical area. The first recorded "mobile" (please note not a "cellular" mobile) telephone call is recorded to have been made on June 17, 1946, by a driver in St. Louis, Missouri.[12] By 1948, this wireless telephone service had been extended to a hundred cities and highway stretches in the United States, and was primarily being used by truck fleet operators.

But this "mobile telephony" system was, to put it mildly, very basic and to boot, expensive to make calls. At most three subscribers in any city could make calls at the same time. The service cost was $15 per month and each call would cost as much as 40¢. Furthermore, the consumer side equipment was heavy and not something you could carry around in your hand or on your person.[13]

The first fully automatic mobile phone system to be developed came from the Swedish company Ericsson. In 1956, they came up with the "Mobile Telephone System A" but which weighed an unbelievable 40 kilograms. A lighter version at 9 kilograms was introduced in 1965, but with very few takers the program was completely wound up by 1983.

In 1957, a Soviet electronics engineer from Moscow, Leonard Kupriyanovich, developed a portable mobile phone dubbed the LK-1. This system weighing 3 kilograms could operate up to 30 kilometers away connecting with a base station. But the set up weighed 3 kilograms. By 1958, Kupriyanovich had designed a smaller model that weighed 500 grams, still too heavy and big to be called a "pocket" model.

In 1947, D. H. Ring, a scientist at Bell Labs had proposed a radically new system. It comprised of several low-power transmitters spread around a geographical area as a hexagonal grid system (the cells), with a technique of automatic phone call hand off from "cell" to "cell."[14] This revolutionary concept was, however, well before its time. The operating technology just did not exist at that time. It would only be the late 1960s that Bell Labs could come up with the electronics to make the "cell" concept work. By 1971, an approval from the Federal Communications Commission (FCC) was received for this cellular telephony concept, and a whole new world of telephony had arrived on the scene.

Since the FCC did not want a monopolistic domination of this rapidly emerging new sector of technology, it encouraged the participation of other companies. Motorola took up this challenge and in 1973 demonstrated their new system called the "Dynamic Adaptive Total Area Coverage" (DynaTAC), the world's first portable cellular phone, nicknamed the "brick" and weighing in at 1 kilogram.

Dr. Martin Cooper, the inventor at Motorola for this product, inspired by the *Star Trek* communicator, is generally credited with making the first-ever call on a portable mobile phone on April 3, 1973, in New York, with the call being made to Motorola's rival, AT&T's Bell Labs. Dr. Cooper would later, in 1992, go on to start his own company Array Comm Inc., a wireless technology entity. Array Comm is now owned by Ygomi LLC and sells subsystems and software for the latest generation of mobile telecommunications.

The concept of cellular-based communications was now rapidly catching on. In 1971, Finland had already tried out their "ARP Network," the first public mobile phone network, now described as the Zero Generation of Cellular telephony (0G). In 1979, the Japanese telecom service provider, NTT, introduced the first automated cellular telephony network. The grouping of Nordic companies followed soon thereafter with the Nordic Mobile Telephone System in 1981.

In 1982, an AT&T subsidiary, Advanced Mobile Phone Service Inc. (AMPS) received Permission from the FCC to establish mobile telephony operations. At the same time, the Indian Army, Corps of Signals, made the world's first cellular-based army communications system, "Army Radio Engineered Network" (AREN) covering the whole country. In 1983, the DynaTAC mobile telephone system was deployed by Ameritech in Chicago.

Once again, there were too many formats and technologies competing with each other for the cellular telephony business. Some were older ones based on analog technologies, whereas others had the newer development based on digital technologies. The United States had opted for its own technologies derived from AT&T and Motorola. The European countries had several of their own developments although the Finns had taken the lead with the deployment of their ARP (Auto Radio Puhelin) System. Furthermore, with rapid changes in technology, there were several different generations of cellular mobile technology on the market with the fourth generation being in vogue now in 2013.

Mercifully, the European countries through their European Telecommunications Standards Institute agreed upon a common standard the Groupe Systeme Mobile (GSM) now also called the "Global System for Mobile Communications." For some years, the GSM system vied with the US radio cellular mobile system the "Code Division Multiple Access" (CDMA). Many of the superior technical features of the GSM standard enabled it to become the more accepted one worldwide although several United States and South Korean service providers continue to use the CDMA system.

The first true GSM network, a second generation (2G) one was also set up in Finland in 1991 by the operator Radiolinja (subsequently named "Elisa"). The 2G network had this great new facility where one could send short text messages from one's cellular handsets. This person-to-person facility of "Short Messaging Services" (SMS) or plain "texting," was also introduced first in Finland and has now become a huge global phenomenon.

With improvements in electronics switching technology, it was now possible to introduce high-speed networks for mobile telephony that could also handle multimedia content with ease. These are termed as the "third generation" networks or "3G" for short. The first successful roll out of a 3G network was made in 2001 by the Japanese company NTT DoCoMo. The pioneers in the use of 3G CDMA networks were two South Korean companies, SK Telecom and KTF, and the US operator Monet who launched the service in 2002. Sadly, Monet has now wound up operations although 3G networks have now become somewhat the international standard with developments moving toward fourth (4G) and fifth (5G) generation technologies.

With the availability of new generation cellular telephony networks and with the evolution of latest electronic components and materials technology, the size, weight, design, and capabilities of handsets also went through radical changes. The heavy "bricks" of early mobile telephony were being replaced by a cavalcade of new handsets. In 1989, Motorola introduced their "Microtrac 9800X," the first truly portable phone. In 1992, the Model 3200, the first digital hand-sized mobile telephone was introduced. But such handsets were still something of a novelty. It was the Finnish company Nokia that in 1992 brought out the very first mass-produced GSM handset, the Model 1011, a 2G device with a talk time before battery recharge of 90 minutes. This set weighing at 475 grams was portable but perhaps just a bit bulky to fit into a shirt pocket.

In 1993, the pair of IBM and Bell South decided they would develop a handset that would be way beyond anything seen so far. It was to be an all-in-one, mobile phone–facsimile machine–pager–diary/calendar–alarm clock–calculator–world clock–address book. It could be had with either a QWERTY keyboard or an on-screen one. It was first offered at $899 and became the world's very first smart phone incorporating what in those days was termed a "personal digital assistant" (PDA). This device, the IBM Simon, unfortunately had several software glitches and was perhaps a little too expensive for the time. It sold only about 50,000 pieces before being taken off the market.

In 1996, an interesting company that had already had a succession of owners over a period of time, decided that there was a market for just a PDA, without the telephony capability. In 1996, this company, Palm Computing, then owned by US Robotics brought out its pathbreaking "Palm Pilot" PDA priced at $299. The following year, 3Com bought out this company but ran into a problem with the name as the Pilot Pen Company of Japan threatened legal action.

By the year 2000, the company was spun off and Palm went public. But by then there had been major improvements in the capabilities of mobile phone handsets and other devices from companies in different countries. As a result, Palm was again in serious trouble. It did try to make a comeback with a smart phone, the TREO 755p in 1997 but by then it was too late. In April 2010, HP bought out Palm for $1.2 billion.

By 1996, Nokia was making rapid strides and was introducing new models of handsets at regular intervals. There was, for example, the "banana" phone, the Model 8110. A "cult" phone because it had featured in the first "Matrix" movie. But it was their Nokia 9000 "Communicator," which would be one of the first true "smart" devices, that would kick start a trend for smart phones ultimately resulting in the demise of Palm.

The Nokia Communicator in a clamshell design, could in addition to telephony and SMS, be used as a more modern PDA. Furthermore, it had capabilities of using the rapidly growing services of the Internet (see chap. 8), not just the "email" function but also the other facilities on offer including browsing. In 1998, Nokia introduced a lighter and more versatile version of the Communicator—this one with a color screen, the Model 9110i.

Other Nokia models would feature diminishing weights and sizes, interchangeable casings, predictive texting, internal antennae, QWERTY keyboards, radio, low-cost GPRS Internet facilities, games, and perhaps the most popular, the camera capability. At one time, it used to be said that Nokia had become the world's largest camera manufacturer. Nokia was for many years the world's largest manufacturer of mobile handsets in the world with a commanding share of the global market.

Yet, somewhere around 2005 things started to not go too well for them. Nimble-footed competitors, with a better feel of the market launched products that were more in sync with what the young of the world, on the one hand, and businessmen on the move, on the other hand, wanted. Nokia's market share began to decline rapidly and since 2010 the company, one of the great electronics companies of all time, was striving hard just to stay in reckoning. Just at the time this book goes into printing in September 2013, the "inevitable" has been announced. Microsoft the software giant, and increasingly a major player in the hardware business especially with its very successful electronic gaming system Xbox, has bought out Nokia for about $7 billion.

While Motorola from the United States was clearly a pioneer in the mobile handset business, barring the solitary effort of the "Simon" by IBM, there were quite insignificant other contributions coming from US companies at that time relating to the manufacture of mobile phones. But the Americas were to have a most surprising challenger to what was rapidly becoming a race between the European and Japanese.

In 1984, two young students, Mike Lazaridis and Douglas Fregin, from Canadian universities set up an electronics design company called Research in Motion (RIM), better known by its brand, "Blackberry." By 1988, it would become the first wireless data communications technology developer in North America and the first company outside Scandinavia to develop connectivity products for wireless data communications networks. Over the following years, the company introduced new products and solutions including radio and wireless modems until the year 1998 when it launched its RIM 950, wireless handheld set.[15]

The next year, in 1999, the company announced its "Blackberry wireless email solution," and had signed agreements with several of the US service providers. By year 2000, it had released its newest handheld, the Model RIM 957 and had tied up with the likes of AT&T Wireless, Bell Mobility, Microcell, CellNet, Nortel, and several others. But the best was yet to come. In 2003, it introduced some really top-class models for the professional consumer including models that could deliver email and Internet capabilities with ease literally on the go.[16] More profoundly, it had extraordinary security software, something big corporates just loved, but the state security agencies just hated. Unfortunately, much as in the case of Nokia, it sort of stumbled against nimbler competition and was attempting a comeback early in 2013. The company would also change its name to be the same as that of their brand, Blackberry, to make it more contemporary.

Needless to say, there were several other companies trying to get market share in the mobile handset business. There were the internationals like HTC from Taiwan, LG from South Korea, and Kyocera and Panasonic from Japan. Sony from Japan and Ericsson from Sweden would then make a partnership. Of course, there were also the several Chinese manufacturers some who were quite happy providing kits for assembly by local manufacturers in different countries. Most of these would survive solely on the basis of price. Many models being no more than copies of those introduced by the market leaders, Nokia, Motorola, Blackberry, and others. The world somehow assumed that this state of affairs would continue for some time. But were they in for a surprise!

The first company to break the sort of status quo in the handset market was a company from South Korea, long known for good-quality, competitively priced television sets and other consumer durables. Samsung was set up in 1938 as an export trading company. In 1958, the company diversified into other lines of business including chemicals, shipbuilding, media, and so on. It was only in 1969 that Samsung ventured into electronic products including radios and television sets.[17]

Samsung's first effort at mobile telephony was a car radio in 1986, but this was a failure as Motorola sets that dominated the market were decidedly better in looks and performance. Samsung at one time almost decided to get out of the mobile business altogether; however, they made another try and launched the Model SH-100 in 1988, the first Korean-made handset. This again did not do well in the market. The years 1993 and 1994 represented a sort of turn around with Samsung's new offering, SH-700, doing quite well in the domestic Korean market.

In 1996, Samsung made its first CDMA system handset and also entered into an agreement to export its handsets to the US operator Sprint. Now, there was no looking back. Exports followed to other countries, and by 1999, the company was the global market leader in the CDMA handset segment. It's first offerings in the GSM mobile phones segment were, however, not too successful. However, by sheer determination, they improved their offerings to suit European tastes and in a few years took away market share from Nokia in the more competitively priced ranges.

The big breakthrough for Samsung in the mobile phone handset business would come in the year 2009. In June of that year, they launched their "Galaxy" series, the model i7500 smart phone using a newly released open source operating system called "Android" from the Internet search company, Google (see chap. 8). The specific Android version used was called the "Donut," which supported large numbers of applications including music and facilitated good-quality still and moving photography. Samsung now had product offerings covering almost the entire range of handsets, from the economical models up to feature-loaded smart phones.

From June 2009 through April 2011, Samsung brought out some 11 new versions of their Galaxy smart phone with updated versions of the Android operating system. By now, Samsung was quite a rage in the market and the leaders such as Nokia, Blackberry, Sony-Ericsson, and Motorola saw rapid declines in their market shares. In April 2011, the company launched its first blockbuster the "Galaxy SII" (Android "Gingerbread") and in May 2012 their other best seller, the "Galaxy SIII" (Android "Ice Cream Sandwich"). By now, Samsung was pretty much top of the heap, with only one challenger—the mighty Apple who had come into the handset business almost from nowhere!

We have read earlier in this chapter how Apple had a smash hit with its iPod range. Until 2001, they were quite busy on that program by which time companies such as Nokia, Motorola, and others including Samsung were well ahead in their mobile phone handset design and manufacturing programs. Late in the year 2000, an Apple engineer, John Casey is believed to have sent out an internal company mail with a concept sketch of a combination of iPod and telephone, which he inevitably called the "Telepod."

Although the concept of the "Telepod" was not immediately carried forward, Apple did make an alliance with Motorola in 2005 to bring out the ROKR E1 handset, which was the first instrument to carry the Apple iTunes collection, albeit only a limited number. Steve Jobs was not terribly happy about this cooperation as it compromised many of the features that Apple's design team under Jonathan Ive would have liked to see on the set. In September 2006, Apple withdrew from this project with Motorola.

In January 2007, Steve Jobs made the announcement that would drastically change the global mobile phone market. He introduced the Apple "iPhone," which he described as "a wide screen I-POD with hand controls...a revolutionary mobile phone...and a breakthrough internet communications device."[18] This small handheld phone, working on Apple's own operating system was based around a revolutionary touch-based user interface operable by a single button along with finger gestures. It had, in addition to telephony, facilities for texting, maps, iPod music, full browsing of an Internet web page, as well as the usual applications of a calendar, clock, address book, calculator, and so on. Pretty much anything one would want in a handheld instrument![19] Apple then went on to announce other versions of the iconic iPhone in the following years, ending with the most recent one, the iPhone 5, announced after the tragic demise of Steve Jobs.

Tablet Computers (Tablets)

The latest handheld gizmo, the "tablet," is not really new. Some developments date even before the following eerily realistic depiction in Arthur C. Clarke's famous science fiction work, *2001 A Space Odyssey*, written, believe it or not, in 1968!

> When he (Floyd) tired of official reports and memoranda, he would plug his foolscap sized NEWSPAD into the ship's information circuit and scan the latest reports from Earth...Floyd sometimes wondered if the NEWSPAD and the fantastic technology behind it was the last word in man's quest for perfect communications.

We did, of course, have Elisha Gray's "Teleautograph" in 1888 and H. P. Goldberg's handwriting recognition interface with a stylus in 1955. But these scarcely qualify under the modern-day definition of a tablet, which says that it is a portable computer contained in a single panel that uses a touch screen as its primary input device.[20]

The "Dynabook" computer for children proposed by Allan Curtis Kay from the University of Utah, came close to the above definition except that the input was by a keyboard. Also, unfortunately, it was never actually built, but if it had it would be like a modern-day "netbook."

Percept Communications Intelligence Corporation in the 1980s developed a pen-stylus-based computer based on handwriting recognition.

We have already read earlier in this chapter about Palm and its range of PDAs, morphing at a later date to be able to do telephony but not quite tablets of our definition as given above. One that came very close was a British company called Psion (short for Potter Scientific Instruments). This company, established by David Potter in 1980, was in the business of developing software and games for Sinclair computers.

In 1984, the company developed its own handheld computer the "Psion Organiser" and an improved version in 1986, the "Psion Organiser II." This one had a touch pad and at least for some of us old fans of Psion, possibly the first "near" tablet by the modern definition. Psion was later acquired by Motorola but not before it had given rise to one of the most significant of all software developments for the mobile phone ecosystem—Symbian—in conjunction with Nokia, Motorola, and Ericsson. Also significantly, it was Psion that first registered the trademark "Netbook."

Apple had been working toward a modern tablet of their own since 1979 when they developed the "Apple Graphics Tablet" a device to create drawings and pictorial inputs into an Apple computer. The design concepts of this device found their way into several of their computer systems.

Steve Jobs persisted with getting further developments done on liquid-crystal flat-panel displays right until he was ejected from Apple in 1985. In 1987, the company had set up a project group to work toward a PDA dubbed the "Newton." First, a tablet-type computer, the P2, was put out as a concept. This was based on Allan Kay's Dynabook. Subsequently, in 1993 following a series of intermediate design stages, a prototype version of the Newton was indeed introduced. However, since this was detracting from the mainstream Macintosh computer program, work on the Newton was put into cold storage.

But it was only in 1989 that an authentic tablet-type portable computer was commercially introduced in the market. This was the "GriDPad" from a company called GRiD Systems Corporation in California, then a part of Tandy Corporation. Unfortunately, they could not make this product a success and the company ended up being sold to Samsung shortly thereafter.

In 1994, that great little British company, Acorn Computers, which started at Cambridge University, was also doing work on tablets. This is the company that gave the world the famous ARM (Acorn RISC Machine), a reduced instruction set architecture that would dominate the global electronics scene for years (see chap. 9). Acorn developed the "OMI News Pad" a touch screen tablet computer as part of a pilot project for the European Union. Unfortunately, by then due to financial difficulties, Acorn had already sold out to Olivetti.

Intel, the famous company manufacturing microprocessors and other ICs, started a tablet program of their own in 1999. It featured an ARM processor, an MP3 player and had Internet capabilities. It was named the "Web Pad." It was a wireless device that shared the home PC's Internet connection and could be operated from any room in the house. It was formally announced in 2001 but failed to get too much commercial traction.

By 1999, Microsoft had teamed up with Xerox's PARC to work on a project to make what was termed the "Microsoft Tablet PC." In 2001, Bill Gates of Microsoft demonstrated in public the first prototype of this tablet PC and launched the device in 2002. Many other companies now came up with their own versions. There was the Nokia 770 Internet Tablet in 2005 and Samsung made one of their own in 2006.

Many others also brought out tablet-style computers. These included Axiotron's "ModBook," Fujitsu's "Stylistic," Zenith's "Cruisepad," 3Com's "Audrey," Palm "Foleo," and more. Unfortunately, all these tablet computers had some fundamental shortcomings. Most were too large to be called handhelds. They also weighed quite a bit and were difficult to carry around. Some suffered from a lack of adequate memory, others on speed. It appeared then that this segment of computers would never take off. But one had not reckoned with Apple.

We know that prior to Steve Jobs's ouster from Apple in 1985 he was very keen on the graphics tablet as also on the concept of a touch screen liquid-crystal-based, flat-panel, input interface. The company's Newton program was to be the one to carry this concept further. Jobs returned to Apple in December 1996 first as an advisor and subsequently becoming the chief executive again.

From around 2003, there had been informed gossip that Apple was working on some kind of a handheld computer. This rumor was fueled by a Taiwanese company (Quanta) manager loosely talking about a large forthcoming assembly contract for a display on behalf of a US major. In 2004, Apple filed for a European patent for a handheld computer listing Steve Jobs and Jonathan Ive as applicants. This was followed up by an application for a US patent in 2005 for the "design" part of the same handheld computer. The drawing in this application seemed to show a person touching a screen with his index finger. The world now had just a vague idea that something quite revolutionary was afoot![21]

It was, however, a new US patent filing in 2008 that made it somewhat clearer that Apple was indeed developing a tablet-type handheld computer with a touch screen that would have multiple finger inputs and other features such as a virtual keyboard and the enlarging of control elements and zooming all by finger touch.

Finally, after years of speculation, in January 2010, Apple announced the launch of its iPad tablet computer. It was a nearly 10-inch backlit touch-type gadget. It included redesigned versions of all the applications contained in the iPhone. It also had many features incorporated from the Macintosh computers but most significantly an entirely new way of accessing iTunes and for reading books via an iBook interface.[22] Many new versions of the iPad would follow in subsequent years. What Apple had created was a totally new market vertical, with many features far superior than either mobile smart phones or small computers, all-in-one easy to use touch interface, handheld device.

Inevitably, the launch of the iPad spurred several other companies to launch competing products. Dell introduced the "Streak," Motorola its "Xoom," RIM (Blackberry) its "Polybook," and many more companies launching similar products. None could match up to the iPad, until that is, Google stepped up to the plate.

This story starts in 2003, when a group of friends in Palo Alto, California, started a company called Android Inc. for developing mobile telephony software. The company had some financial problems, so in 2005 they sold out to the Internet search software firm, Google, which was keen to get a piece of the action in the burgeoning "mobile" field.

Under the guidance of Google, the founding team of Android, which had stayed put, developed an excellent mobile device software (the Android system) based on the open source platform "Linux." The open source platform meant that the base software could be tweaked to the specific requirements of individual hardware manufacturers and also would be freely upgradable. For Google, the benefits would come out of the large numbers of manufacturers who would sign on for this innovative software and produce very many devices, all using Google's ubiquitous "Search" function.

In 2007, a consortium of leading companies, including Samsung, Sprint, Qualcomm, TI, and others established the "Open Handset Alliance." The first "product" out of this alliance was the "Android" operating system. The following year mobile phones using this new operating system started to be introduced in the market and soon this became a veritable flood with numerous handset manufacturers, including as we have noted, Samsung, now signed on to use the Android operating system.

By 2009, the Android operating system had been tweaked to be used in the tablet computers and other small computers of the time. Among the early users were the Archos Tablet, Acer Net Book, Dell, ASUS, and several others. But it was only in 2010, shortly after the iPad had been launched that Samsung that had already been hugely successful with its range of Galaxy handsets, decided to get into Android-based tablet computers in a big way.

The first Samsung "Galaxy Tab" came on the market in the year 2010. Since then, there have been newer models at varying price points. In a very short time, Samsung had taken away market share from the dominant Apple in many countries of the world. Suits and counter suits relating to technology and design infringements soon followed as both market leaders fought to be the "Number 1," but then that is another story altogether.

So where are we now headed in this whole business of mobile phones, smart phones, and tablets? The numbers are already mind boggling. There are over one billion smart phones already in use worldwide. The numbers of mobile phone subscribers is now over six billion and increasing quite literally by the hour. The numbers of tablet computers sold annually in the world now exceeds 260 million and continues to rise as prices decline and newer markets emerge. Unfortunately though, tablets are becoming so ubiquitous that toddlers at the age of 4 and 5 are getting addicted to their use. Parents use the device as a surrogate baby sitter. Not so incredible then are recent reports that the Capio Clinic in London now runs a "digital detox" therapy program for little children!

Companies will change leadership positions. Some may lose ground as has happened with the likes of Nokia and Blackberry. There will be other companies who will mess up their markets as Motorola did in China with advertising campaigns that defied all cross-cultural sensitivities. Some names may even become history or others like Google, a great search engine software company becoming a hardware manufacturer by taking over Motorola Mobility.

As we write, we know that technological changes are forthcoming. Smart phones and tablet computers are merging in some models to become another market segment, the "phablet." We also learn that in the very near future there may be another new communication device based around the wristwatch. Dick Tracy be praised! Another class of devices may even use eyeglasses for communications as well as entertainment. There will inevitably also be newer operating systems emerging such as "Sailfish" from Jolla (Finland) and "Ubuntu" from Canonical (UK). The future looks extremely interesting indeed!

CHAPTER 8

Computer Networks and the Internet

Governments of the Industrial World, you weary giants of flesh and steel, I come from Cyberspace, the new home of Mind.

—John Perry, lyricist of "The Grateful Dead"

Those who have read Arthur C. Clarke's famous short story, "Dial F for Frankenstein," published in 1961, may recall his prescient prediction of an increasingly interconnected telephone network that sort of goes berserk and starts global chaos by taking over all financial, transportation, and military systems. So Clarke, much as he had also visualized a tablet computer, did have visions of an interconnected communications system.

True, the possibility of getting computers into some kind of network had been visualized earlier. In 1945, the great American scientist and engineer, Dr. Vannevar Bush, had written an essay titled, "As We May Think." In this essay, he describes a theoretical machine, the "Memex," in which one could store and retrieve documents linked by associations, what in present-day terminology is known as "hypertext."

But blame it on the Russians! Virtual panic set in when they launched their satellite, the "Sputnik" in 1957, beating the United States into space. The Western world was left in considerable shock! Remember that there was a virtual state of cold war on at that time. The United States in particular, now needed to pool all its scientific and technological resources to ensure that not only they did not fall too behind the Soviets but actually in defense related matters needed to be well ahead.

To begin with, a new entity needed to be formed that would be the driving force behind the development of the most-modern technologies especially those related to defense matters. Thus in 1958, the Advanced Research Projects Agency (ARPA) was created. ARPA's remit was to "formulate and execute R&D projects that would expand the frontiers of technology beyond the immediate and specific requirements of the Military Services and their Laboratories."[1] The name of this organization would later be changed to the Defense Advanced Research Projects Agency (DARPA).

It was widely believed in the US scientific community at that time that there were many technologies and other scientific knowledge available in different institutions geographically spread around the country but there was very limited information exchanged between the various institutions and almost none on a real-time basis. Further, there were certainly no real-time linkages between repositories of scientific and technological knowledge in the United States with the institutions of friendly countries.

In 1960, Dr. J. C. R. Licklider, who had earlier worked with the MIT as an associate professor on the team developing a real-time air defense computer, wrote a paper on "Man Computer Symbiosis" in which he postulated that computers should be developed with the goal, to quote Licklider, "to enable men and computers to cooperate in making decisions and controlling complex situations without inflexible dependence on predetermined programs." In essence, this was all about real-time interactive computing.

In 1961, Dr. Fernando Corbato of the MIT Computational Center developed the "Compatible Time Sharing System" (CTSS), which most people regard as presaging what would become the "electronic mail" (email). The CTSS system, however, would get operational later in 1962. In 1961, Leonard Kleinrock, a graduate student at MIT wrote his thesis on "Information Flow in Large Communication Networks," describing the theory of packet networks, a technology that is the basis of what would subsequently become the Internet. Interestingly, it was Dr. Kleinrock's computer at University of California, Los Angeles (UCLA), that became the first node of the Internet in September 1969, as we will read later in this chapter.

In 1962, the RAND Corporation was tasked to work on a project to provide command, control, and communications to US bombers and missile bases in the event of an enemy nuclear strike. The results of this work clearly showed the need for a decentralized ecosystem of electronic "packet" switches. The Santa Monica–based System Development Corporation (SDC) evolved out of RAND. In 1965, on the basis of the earlier work at RAND, SDC developed a messaging system for the Strategic Air Command running on decentralized switches. This was clearly a precursor to electronic mail.

Late in 1968, ARPA awarded a one-year contract to a company Bolt, Beranek, and Newman Inc. (BBN) for the development of the essentials of a network of four Interface Message Processors. The design was built around off-the-shelf Honeywell computers and comprised four nodes, the first at UCLA, the second at Stanford Research Institute (SRI), third at University of California, Santa Barbara, and the fourth node at University of Utah. On October 1, 1969, the first successful transmission of data was carried out from UCLA to SRI. The ARPANET computer network system was now born. Other computer nodes were added and by April 1972, the network had been expanded to a total of 23 nodes.[2]

The ARPANET node at the SRI was at the "Augmentation Research Center," an ARPA-funded entity doing research work on defense-related projects. This entity had been established by a somewhat reticent computer "wonk" or "geek" named Douglas Engelbart. Engelbart strongly believed that the future of technological development lay in scientists, engineers, researchers, and academia at large being able interactively to create, share, work on documents, and other sources of information in real time from different geographical locations. Engelbart designed just such a system, which he called the "oNLine System," NLS for short.

But Engelbart was to become really famous for designing a small piece of computer hardware that is so ubiquitous and widely used today that not many give it a second thought. In the early 1960s, getting work done on the computers of the time required piles of punched cards, which would be used for information input. One then waited patiently for the information to be processed and results churned out.

In 1968, Engelbart stunned the world by demonstrating to a live audience at an International Computer Conference, a "gizmo," comprising a small wooden box on rollers, connected by a length of wire to a computer. By using this gadget controlling a "cursor," he could open windows, give commands for text editing, and also set up real-time interaction with another user some distance away. What he had developed was the computer "mouse."

The technology was further developed at the PARC and licensed out to Apple. Sadly, Engelbart died in July 2013 without quite having gained any real monetary benefits or global honors for having invented the computer mouse.

In 1973, Robert Metcalf, a research engineer at the Xerox's PARC, was asked to work on a system to interconnect the "Alto" computers that PARC had developed (see chap. 6), and were being used in the center, as also some other makes of computers. There were many numbers of these computers and more importantly, the users of these needed to be able to get prints from their new "play toy," the laser printer (also developed at PARC). By 1976, Metcalf had developed a very efficient system for a local area network that he dubbed the "Ethernet" and received a patent for it. Today Ethernet has become an industry standard and the most widely used system in the world for a local area network of computers.

Electronic Mail (Email)

Prior to the ARPANET, electronic mail messages could only be sent to users of the same computer system. With an interconnected network of computers in different geographies, there was the need to put addresses of specific intended recipients of electronic messages. Ray Tomlinson, a principal scientist at BBN had already worked on some mailing software. In late 1971, he decided to combine some of his earlier work into a comprehensive file transfer program and later sent out the first recorded electronic mails (emails) on a network, and used the "@" symbol for separating the addressee's name from that of his computer.[3] In 1990, BBN was bought out by Raytheon.

As has been the case throughout the history of electronics, even email has had its own bit of controversy. While Tomlinson is widely credited with its invention, there is another lot of people who believe that the credit for this must go to Shiva Ayyadurai, a computer savvy young student of Indian origin whose mother worked at the University of Medicine and Dentistry, New Jersey. In 1978, it is said, Shiva Ayyadurai was asked by a teacher of this university to create an electronic system for various interoffice communications being sent around the campus. The electronic system was required to have all the normal requirements including "To," "From," "CC," "Subject," and so on, along with Inbox, Outbox, and folder capabilities.

Such a system was indeed developed and introduced in 1980 and a US copyright was issued in the name of Shiva Ayyadurai in 1982, albeit only for his email software product and not for the technology per se. Shiva Annadurai went on to study at MIT, received a doctorate and subsequently became an entrepreneur in the United States, promoting companies including the multimillion-dollar Echo Mail.[4]

In 1973, Vincent Cerf at Stanford and Bob Kahn at DARPA (the renamed ARPA) started to work on software protocols that would enable computers on different networks to communicate with each other. The result of their endeavors was the development of the "Transmission Control Protocol" (TCP), which forms the core of the "Internet Protocol" (IP). In 1974, Cerf and Kahn would go on to coin the term "Internet" for the first time.

Suddenly, various "nets" started to proliferate. The first international connectivity into ARPANET was provided in 1973 to University College of London and the Royal Radar Establishment of Norway. The same year the fundamentals of Ethernet, a local area networking system, were developed by the Xerox's PARC, with help from Intel and DEC. In 1976, a satellite program called SATNET was developed to provide an Internet link between the United States and Europe enabling Queen Elizabeth to send out her first email message in the same year.

By 1981, in the United States, computers were now allowed to network without connecting into the government's own networks. But the next steps of considerable significance would come in 1983. TCP/IP developed by Cerf and Kahn had now become the global standard protocol for Internet and more importantly, the "Domain Name System" (DNS) was introduced that year. DNS made it so much easier to remember email addresses. As a random example, www.web-hosting-top.com is so much easier to remember than the convoluted IP address of 80.121.204.234.

With the introduction of the World Wide Web (see below), electronic mail or email as it is popularly known, became even easier and its use became widespread. Several email service providers and "portals" started offering this service, including, for example, AOL, Yahoo, Google (Gmail), Zoho Mail, and Inbox.com. Several email service providers started out as a paid service but very soon moved to an advertisement-based free service. The most interesting of these would be an entity called "Hotmail."

Hotmail was set up as a "Silicon Valley" start up in 1996 by a young person of Indian origin, Sabeer Bhatia, along with a friend of his, Jack Smith. The name itself derived from the language used for the software, "Hyper Text Mark Up Language" (see below). Providing a free service, the company grew extremely rapidly and in just under a year had more than a million subscribers.[5] Clearly, the company now presented a juicy takeover target. Late in 1997, Microsoft acquired Hotmail (now renamed as "Windows Live Hotmail") for a reported sum of $400 million and merged it with another acquisition, a calendar service called Jump. Presently, the service has over a billion inboxes and several hundreds of million users globally.

By 1988, the Internet had indeed become a great means of computer networking, data transfer as well as electronic mailing. But its use was predominantly at the level of universities, research establishments, large enterprises, and the government. In 1989, the ARPANET having served its purpose, ceased to exist. A further development to enable the common person to make use of this great innovation—the Internet, was now needed.

World Wide Web and "Browsers"

The development, a radical one, came in 1989. Tim Berners-Lee, a British scientist was working at CERN (the European Laboratory for Particle Physics) in Meyrin, Switzerland, now famous for its Large Hadron Collider and the discovery of the "God" particle. Berners-Lee made a proposal to merge the technologies of PCs, computer networking, and hypertext (text, pictures, and programs creatively linked to each other) all into one easily usable global information system that could be viewed by special software programs called "browsers." He called this the "World Wide Web" (WWW).[6]

By the end of 1990, a prototype "web" system with a computer serving as the sort of anchor node (the "server") to communicate with was demonstrated and the world's first web page address http://info.cern.ch/hypertext/WWW/TheProject.html was launched. Shortly thereafter, the web system extended for use not only by the various departments at CERN but also by scientists at other scientific establishments.

Time and other resources at CERN were clearly not adequate to spread the gospel and increase the usage of the World Wide Web. Tim Berners-Lee launched a plea by Internet for other software developers to join in the effort.[7] The most significant contribution came in from a group of four students from Helsinki, Finland. They started to develop a user friendly browser in 1991 and had it completed by April 1992. They called their product "Erwise," which became the world's very first "graphical browser."

Unfortunately for the poor Finnish students, their country at that time was going through a deep economic depression. There just wasn't enough venture capital available and no other commercial backers were forthcoming. The possibility of another blockbuster entity from Finland to follow in the footsteps of Nokia had now completely receded.

While the young Finnish students were almost finishing work on their browser, another student, this time one from Taiwan studying at the University of California, Berkeley, Pei-Yuan Wei, wrote a program called "Visually Interactive Object Oriented Language and Application" (VIOLA, for short), in April 1992. This became the first browser with inline graphics and scripting capabilities but suffered from its inability to be ported onto a PC.

Two other browsers then came on the scene. The first was called "Midas" and was developed by Tony Johnson at the Stanford Linear Accelerator Center (SLAC) and released in the summer of 1992. The use of this browser was largely intended to be done by scientists at SLAC, which by then had a working relationship with CERN. The second was a browser, called "Samba" developed at CERN itself for porting onto the Apple Macintosh computer and was by and large a derivative of Tim Berners-Lee's original work.

The first browser, to really catch on with Internet users at large, however, came from the National Center for Supercomputing Applications (NCSA) based at the University of Illinois, at Champaign–Urbana. The NCSA was established in 1986 as one of the sites for the National Science Foundation's supercomputer program. In 1993, a team from NCSA developed the "Mosaic" browser with some unique features such as icons, bookmarks, and pictures. Further, the browser was available to be taken free of charge directly from the NCSA website. Within a matter of days, many thousands were being downloaded each days.[8]

The Mosaic technology suddenly became much sought after now. More than hundred licenses were given out including to a start-up company named Spyglass founded by Tim Krauskopf as well as to a company called Mosaic Communications Corporation (subsequently renamed as Netscape), set up by Jim Clark (founder of Silicon Graphics) and Marc Andreessen who had earlier worked on the Mosaic project at NCSA.

Spyglass was an Illinois company, founded in 1990. They were working on scientific data analysis tools before they decided to seek a license for the Mosaic browser from NCSA. They even obtained the rights to the Mosaic trademark. Their business model for the browser was to develop the basic browser technology and then sell to corporate intermediaries who in turn would deliver a fully packaged product to end users.

By 1994, Spyglass had almost 120 licensees for the "Spyglass Mosaic," one of them being the redoubtable Microsoft who used this browser in their "Windows" platform in 1995 and called it "Internet Explorer."[9] Spyglass afterward alleged that Microsoft had more or less hounded out all the other licensees and further, that having bundled the browser into the Windows platform, Microsoft paid only the minimum royalty fees and not on the numbers of Windows licenses incorporating the browser. The companies then went into a legal dispute that was settled by a financial payout by Microsoft. All further developments on the Internet Explorer browser were then done by Microsoft who brought to bear considerable resources on this program. Spyglass would end up being acquired by OpenTV in the year 2000.

Netscape, with the backing of Jim Clark of Silicon Graphics, was considerably better funded than Spyglass and hence could bring to bear substantially more resources onto their browser project. The Netscape team worked at a furious pace as they realized that Spyglass had started much earlier and also was in advanced talks with Microsoft. Before the end of 1994, Netscape had released its first browser, which immediately caught on in the market and at one time even had an 80 percent market share. By 1995, the company went in for a public issue. It was to be at that time NASDAQ's third largest ever initial public offer by share value.

Over the next few years, Netscape introduced newer versions, the most significant of whom would be the "Netscape Navigator" in October 1996, which had the ability of handling the "Java" language. But by now Internet Explorer was becoming a major success and taking away a significant market share. Netscape tried to counter by announcing that their next release would be free and be based on open source technology. Unfortunately, this new project called Mozilla got delayed by many years. Microsoft's Internet Explorer was now way ahead. By 1998, Netscape had sold out to America On Line (AOL).

AOL itself was acquired by Time Warner in the year 2000. The timing was horrendously wrong as the whole industry was in turmoil, as we will read later in this chapter. The $165 billion acquisition has been described as possibly the greatest mistake in corporate history. In 2009, AOL was spun out as a separate entity. Even by 2013, it is still to put its head above water!

There have been and still are several other Internet browsers around, but three need special mention here. One has come out of Norway, unbelievably. Sometime in the middle of 1994, two engineers at Telenor, the Norwegian telecommunications company, decided to develop a really fast browser for computers that may not necessarily have top-of-the-line features. The duo decided to work from scratch and did not follow the basic Mosaic browser template. By 1995, they had developed the "Opera" browser with some exceptional features. Opera would rapidly go on to acquire a "cult" status and continues to do so with their frequently updated releases.

Another browser, Firefox, was developed by the Mozilla Corporation. This followed on from work done earlier at Netscape before the takeover by AOL. This browser was released in the market in 2004 and being an open source product, has acquired a considerable market share.

The other browser, inevitably came from the Internet search company, Google. A few software professionals who had earlier worked on the Mozilla browser development joined Google in 2007. By 2008, they had developed the "Chrome" browser, which was then released in several different languages of the world.

Despite the availability of several good alternatives, Microsoft's Internet Explorer continues to dominate the market, so much so that the

US government filed an Antitrust Law Case and the European Union also followed suit. As a result, the company had to come to a settlement in both cases and essentially offer their browser to users unbundled from their Windows operating system.

The Growth of the Internet—the Search Engines

At the time, Netscape launched its browser in 1995 and more so since it went public in 1996, the Internet and associated businesses were on an unprecedented growth path. Everyone now wanted to be a part of this new phenomenon, including the food company, Pizza Hut, who introduced the first Internet-based ordering system. There was even a start-up Internet bank, the world's first virtual bank.

Several software and other companies, in many countries started to provide services for website designing and other Internet-related inputs for what was rapidly becoming an Internet-based world. Since most companies and entities had got on to the World Wide Web wagon where most of the unique resource locators ended in ".com," one may say that the world had now rapidly entered a "dot-com" ecosystem.

Of course, there were many software and service companies who wanted to capitalize on this new dot-com phenomenon. They went ahead and announced public shares issues at unrealistic pricing, seeking "valuations" on the basis of dreamt-up future earning potentials without any underlying assets or visible cash flows. Punters and venture capital companies in many countries started putting out good money to invest in these Internet companies. In March 2000, the NASDAQ composite index touched the high of 5,100. Something had to give way, and it did! By 2001, the stock markets had a major correction. Many dot-com companies saw a huge drop in their share value. Many ceased to exist all together. The dot-com bubble had burst!

The growth of Internet businesses after the bubble had burst was painfully slow and measured and in large part relied on a few businesses that had managed to survive the crisis. Some of these were what started to be called "e-commerce" companies that used the Internet to buy and sell goods and services. The two main ones in this category were Amazon (founded in 1994), that started life selling books on the Internet before adding movies, music, electronics, other fast moving items, and so on. The other was eBay, a company set up in 1995, for auctions and retail sales on the Internet.

With the growth of the Internet, world now on a sounder footing, the whole ecosystem turned on to a rapid growth path again. Companies around the world began to see the advantages of having information about themselves and their products and services readily available to existing and prospective customers. New websites and web pages multiplied rapidly.

By 1998, there were some 26 million pages on the Internet. In a matter of two years, these had grown to a figure beyond a billion and by 2008 the number exceeded 50 billion. It is recorded that by the year 2008 there were already in excess of 170 million web addresses (unique resource locators). Clearly, there was an information overload on the Internet. The common Internet user needed an easy way of looking for specific information of interest. Enter the "Internet Browser."

The first Internet "search" engine of sorts was the one developed at McGill University in Canada. As early as 1990, the university had announced the development of "Archie" (a shortened form of "Archive"). But Archie only had an index of directory listings, and not the contents for each website, due to limitations of space. In 1991, somewhat improved versions were introduced. They were called "Veronica" and "Jughead." These operated much like Archie except that it was now possible to use plain text files. Further, they predominantly used text files in a format called "Gopher" developed at the University of Minnesota in 1991.

There was, of course, Tim Berners-Lee's own "Virtual Library" (ViLib) that was available for use by 1992. Some others followed in 1993 including "Excite," "World Wide Web Wanderer," "Aliweb," "Primitive Web Search," but neither of them quite caught on. The year 1994 was to be much better. First, there was "Infoseek," which could give up a page on the World Wide Web in real time and was used by Netscape as a default search engine. "Altavista" followed shortly thereafter and offered for the first time natural language search questions as well as an unlimited bandwidth. Another search engine was "Lycos," which ranked searches by relevance but Lycos was shortly sold off to Daum Communications of South Korea.

Later in 1994, two students from Stanford University, Jerry Wang and David Filo, set up a company called Yahoo, initially named "Jerry and David's Guide to the World Wide Web," for bookmarking and listing interesting websites. Officially, the name of the company is an acronym of "Yet another hierarchical officious Oracle," although the slang version is perhaps more appropriate. Before the end of the year, they had already registered more than a hundred thousand unique visitors.[10]

The following year, in 1995, Yahoo received venture capital financing and rapidly developed into the world's first recognized Internet brand after having developed the pioneering navigational guide to the

Internet. Until 1998, Yahoo was extremely successful, but when the dot-com bubble burst it had a major effect on the value of Yahoo's stock. From 2002, the company started acquiring other search engine companies and also started to work in close cooperation with many industry leaders expanding on their Internet-based offerings.

However, competition was not only catching up but actually was also leaving Yahoo considerably behind. For several years, starting in 2005, Microsoft made attempts to buy out Yahoo but they were rebuffed each time. However, in 2009, a deal was announced whereby for a period of ten years Microsoft would be allowed to use Yahoo's search software as a part of their own search engine, called "Bing."

It was 1996 when the "daddy" of all search engines was developed and released. The previous year, that is, 1995, two students, Larry Page and Sergey Brin, had met, believe it or not, again at Stanford University. The following year, after graduating in computer sciences the pair began working on a search engine. They called it "Back Rub" and operated it from the computers at the university. The following year the two decided that the name needed to be changed and came up with "Google" derived from the word "Gogol" denoting the numeral "1" followed by hundred zeros.[11]

In 1998, the cofounder of Sun Microsystems, Sandy Bechtolsheim (see below), sent out a cheque for the amount of $100,000 which was deposited only after the company was formally incorporated a month later and a bank account opened. The company started its operations from a garage in Menlo Park, California. The year after, in 1999, the company received $25 million in venture capital funding. Google was now well on its way.[12]

By the year 2000, Google had launched its search engine in 15 languages and reached the milestone of a billion website addresses, making their search engine by far the largest in the world. Also that year the company entered into a deal, making their browser as the default one on Yahoo. Further, by 2002, the company entered into agreements to start serving 34 million customers of Compuserve, Netscape, and AOL. Google had now become a generic word for Internet search, much as happened with "Xerox" for photocopying or "Frig" for a refrigerator.

Whatever information one wanted, meanings of words, government regulations, travel information, weather reports, currency conversion rates, and pretty much anything else, one "googled" for it. Promptly, but naturally, the *Urban Dictionary* came up with the word "Ungoogleable," setting off a rather amusing tiff with the somewhat overbearing Swedish language watchdog, which would not allow such a new-fangled word![13]

Google went on to enter other areas such as Internet browsers, email-ing, smart phones and their operating systems, television, books, online photography storage and editing, video handling and streaming pro-grams, and much more. It would rapidly become one of the world's most valuable companies! As this is being written, Google is deeply involved in a project to perfect a "driverless" car, and perhaps has more great ideas in its kitty!

The web search business in April 2013 was estimated to be around $22 billion with Google being the clear market leader with about two-thirds of the market. But the market now appears to be veering toward smarter search options where customers are looking for direct answers and customized results to specific queries. Increasingly, this implies the use of dedicated websites of market leaders in specific verticals.[14]

It is now over two decades since the World Wide Web started. Some like the Internet pioneer Vincent Cerf would like to believe that the Internet is now thirty-odd years old and is now "in a mid-life period." What began as a defense-related project grew into what may now be seen as a hugely disruptive innovation with an extraordinary global impact. In a recent interview, Cerf had the following to say, "The internet era is different from the telephone era for at least two reasons: It allows groups to communicate, co-ordinate, collaborate and share information, and it supports every medium of communication invented, all in one network. People can discover each other without knowing who they are."[15]

The Internet has grown to some 2.5 billion users today from 16 million in 1996. The number of users of the Internet is estimated to cross a mind-boggling 6 billion by 2020. By 2012, the total num-ber of registered domain names (website addresses or unique resource locators) exceeded 250 million. People were now increasingly intercon-nected, and hence the operative term "online." The Boston Consulting Group claims that if the Internet were a country "it would rank as the world's fifth largest economy." It is poised to grow further with the so-called industrial Internet for large companies yet to become prevalent. It took 70 years for telephones to reach 50 percent households, just under 30 years for the radio to achieve that. The Internet did it in only ten years.

The Internet is now used for all kinds of applications such as getting the news (jeopardizing the very existence of newspapers); telephony (the likes of entities such as "Skype," now part of Microsoft, taking away chunks of business from operators); radio (see chap. 3); video and mov-ies; books and magazines in electronic form (which can be read off handheld tablet like devices); e-commerce and shopping (note the rise of the Chinese company Alibaba, now a behemoth with revenues of $4 billion); pornography for those so inclined; electronic gaming; and most recently, social media sites such as Facebook, YouTube, Instagram, Twitter, and Google+ that have the youth of the world hooked possibly to the detriment of their academic and mental growth.

The rapid rise of the Internet, of course, has its dark side. It has been increasingly used for cyber warfare. Several countries actively commit cyber attacks against deemed and perceived enemies. Some now even have trained cyber warrior cadres and are almost beginning to be seen as another arm of the defense services of the land.

There are a huge and growing number of unwanted email messages being sent out, many using machine mailers. Some are just irritating messages termed as "spam," but others are extremely dangerous messages (now called "phishing") and attachments that try to get hold of passwords, confidential information, banking data, and so on from unsuspecting users.

But the really bad side of the Internet is in the deliberate and planned spreading of malicious programs (malware) or viruses either as phishing attacks, state sponsored or terrorist attacks targeting the infrastructure of other states or large enterprises (the "Stuxnet" attack on Iran, for example), or just straightforward nasty maliciousness. The number of attacks from viruses number over a million now, such as "Trojan," "Malicious URL," and many other attacks from other computer "worms," "phishing," "spear phishing," hacking, and malware runs into many hundreds of millions. In the year 2011, the United States reported that some 50 percent of Internet users detected virus threats and Russia almost 56 percent.

All of this, of course, means good business for speciality companies making antivirus software including the likes of Symantec, Kaspersky, and AVG Technologies, but somehow the bad guys always seem to stay one step ahead. Speciality software companies have now also been set up to sell at many thousands of dollars, "exploits" (computer codes to exploit flaws in operating systems so that infiltration into computer systems becomes possible). It is not just the governmental agencies that are buying these exploits. The bad guys are at it too![16]

Networking Hardware

Of course, to do all the wonderful things that one can do on the Internet, as one can imagine, there needs to be a full paraphernalia and ecosystem of hardware and associated software. It is clearly not as simple as dialing a phone number and getting an answer from the called party on the other side.

The vast amounts of data used by Internet service providers, websites, and other enterprises are stored on special computerized systems called servers. Large numbers of these exist at single geographical locations, and are interconnected. This is referred to as a "server farm," "cluster," or sometimes even as a "ranch." Many of these server farms are in very large warehouses and buildings, requiring massive air conditioning and the associated electrical power to run not only the servers and other electronic equipment but also the climate-control facilities. One is even housed in an old NATO facility in Reykjavik, Iceland, a country with vast amounts of spare electrical power from geothermal sources and because of the already cold climate, a minimal need for air conditioning.

Google has set up a large server farm in a disused paper factory in Hamina, Finland. Cold water is drawn in from the Gulf of Finland and passed through a direct heat exchanger, extracting all the waste heat before the hot water is pushed back into the gulf. The $300 million server farm at Hamina has only a handful of employees and the whole place looks surreal, as if from a James Bond movie!

To set up a network of servers, one needs electronic "switches," which act as sort of controllers connecting an assortment of these switches with computers and other devices such as printers. To interconnect networks, however, other electronic equipment called "routers" are used. These electronic boxes determine the optimum route that an electronic signal should take to establish connectivity. Thus, the switches and routers act as fundamental building blocks in setting up modern data communication networks.

There are many companies, from several countries, that produce the electronic systems for the Internet. Many are leading manufacturers of computer systems and other hardware. It would be nigh impossible to name all the manufacturers and to cover their histories in this book. So let us look at a few of the most important of these companies.

Much like several other great electronics companies, Sun Microsystems also grew out of Stanford University. In fact, "Sun" is an acronym for "Stanford University Network." Andreas Bechtolsheim, a graduate student at the university, in 1981, developed a more powerful, single-user computer called the "workstation," for advanced engineering computation using off-the-shelf electronic components. The following year, 1982, Bechtolsheim was joined by two other Stanford graduates, Vinod Khosla and Scott McNealy, to set up the company.

Within a short time, Sun had launched its first products in the market, which became instant successes. The company had sales of $8 million in its very first year. Shortly thereafter, one of the major computer aided design companies, ComputerVision, dumped its own proprietary hardware in favor of workstations built by Sun. The company was now on a roll and between 1985 and 1989 it was deemed to be the fastest growing company in the United States.[17]

Over the next few years, Sun developed and sold even more powerful workstations running on open source software as also a newly developed software called "Java" making Sun the global market leader. With the rapid growth of the Internet, by 1990 the company expanded their market offering to include servers. However, with the bursting of the dot-com bubble in 2001, the company was hit quite hard.

Their top-end and expensive servers suddenly found new competition from low-cost servers designed around powerful semiconductors. It made a partnership with the Japanese company Fujitsu to use their IC processor chips but the glory days of the company were now behind them. In 2010, the company was acquired by the software major, Oracle Corporation.

In 1984, Stanford University would be the source for another great company. A husband and wife team of Len Bosack and Sandy Lerner, computer engineers at the university, teamed up with another computer engineer, Richard Troiano to set up Cisco Systems. The name of the company derived from San Francisco, the lovely big city near Stanford University.

The duo developed hardware (routers) that could interconnect the networks of the two departments that they worked in. Since the technology was developed at Stanford, the university tried to collect a huge licensing fee. Eventually, this fee was settled at $150,000 together with the supply of free routers and services.[18]

Cisco struggled until 1988 as a seller of computer networking products, mainly simple routers. A venture capitalist was soon found for the company. Equity financing was obtained but management control had to be ceded to this venture capitalist. The management team comprising the original owners was also now replaced.

With the Internet usage now really picking up after the dot-com bust, demand for Cisco's products started to grow rapidly. By 1990, many companies were establishing local area networks comprising PCs. By 1992, Cisco had sales in excess of $330 million and had become the second fastest growing company in the United States. Cisco now added newer and better products including servers and switches, the technology for which was obtained by acquiring a company called Kalpana (set up by Vinod Bharadwaj an entrepreneur in the United States of Indian origin). The company also made a significant entry into overseas markets, becoming hugely successful.

In 1995, a new chief executive, John Chambers, joined and raised the company to even newer heights based on some successful acquisitions of existing high-technology companies. By the year 2000, the company's market value had soared to $450 billion making it briefly the third most valuable company in the world after Microsoft and GE.

Over the years, despite increased competition, Cisco has remained, arguably, the dominant global player in the networking-electronics hardware field. It is then little wonder then that recent Cisco print advertisements carry the following paean:

> And tomorrow we'll wake up pretty much everything else you can imagine.
>
> Trees will talk to networks will talk to scientists about climate change.
>
> Traffic lights will talk to cars will talk to road sensors about increasing traffic efficiency.
>
> Ambulances will talk to patient records will talk to doctors about saving lives.
>
> It is a phenomena we call the Internet of Everything.
>
> Tomorrow? We're going to wake the world up. And watch, with eyes wide, as it gets to work.[19]

In 1979, Robert Metcalfe, an engineer with a degree from MIT, quit his job at Xerox's PARC to set up a consulting company providing support services for computers, communications, and compatibility, and hence called this company 3COM. The following year, the company entered the area of developing hardware for interconnecting computers with peripherals such as printers in a local area network. With financial inputs from venture capitalists, 3COM established their own manufacturing for their newly developed products.

With the increasing popularity and use of PCs, 3COM had considerable success, which encouraged it to go in for a stock market flotation in 1984. Two years later, by 1986, the company had revenues in excess of $60 million with a profit margin more than 15 percent. By then, it had also introduced servers in the market, which accounted for 35 percent of total sales. In 1987, it acquired another networking products company, Bridge Communications. The same year the company made an agreement with Microsoft to develop and sell networking software.

With increasing competition and reduced revenues from their software business, 3COM in the 1990s changed tack and became once again a pure hardware company. With the help of several company acquisitions, by 1996, 3COM posted a turnover of over $2.3 billion and making it possibly the second most important company in networking hardware, after Cisco. But this apparent success led to the company making a somewhat disastrous takeover in 1997 of US Robotics, which had acquired Palm Computers, at one time a very successful manufacturer of PDAs. Even a joint venture with the Chinese telecom company, Huawei, giving a good part of market share in China could not turn around the fortunes of the company. By 2009, HP had started the acquisition process of 3COM.

In the year 1996, another former employee of Xerox's PARC decided that he would also go into business for himself to address the burgeoning business opportunities provided by the Internet. By 1996, Pradeep Sindhu, an engineer of Indian origin and a graduate of the Indian Institute of Technology, Kanpur, had set up his company, Juniper Networks, and design work on their offering of a router initiated. In September, the M-40 router was introduced to the market with specifications and performance better than any other similar router.

Juniper went public in 1999 in what was one of the most successful offerings of that time in the United States. The company very rapidly ramped up its product offering and also went on an acquisition spree. From 1999 through 2012, Juniper acquired 18 companies including Micromagic, Unisphere Networks (a subsidiary of Siemens), NetScreen Technologies, SMobile Systems, and others.[20] The company made steady progress and in 2009 was ranked fourth in *Fortune* magazine's list of "Most Admired Companies." Juniper subsequently announced its latest range of switches as also a partnership with IBM for the sale of advanced routers and switches in competition with Cisco.

CHAPTER 9

"Chips" and Displays

Nature abhors the vacuum tube.

—J. R. Pierce, engineer at Bell Labs who
coined the word "transistor"

The more I.C., the less I see.

—Anon

For all the exciting electronics stuff that we have around us to operate properly, the right components are needed inside these products. In the good old days of the technology, when, for example, radios were just coming into vogue, a halfway-decent set needed a set of electronic vacuum tubes or "valves" as they were called. The most common ones were the triodes of de Forrest and Tigerstedt (see chap. 1). The rest of the circuitry involved what are dubbed as passive components (which included resistors, capacitors, coils, etc.).

The use of valves in electronic equipment required the availability of a fairly heavy power supply, which would provide the voltage for operating the valves. The reasonably high voltage required, conventional-valve-based (vacuum tubes)radio sets had to be operated on electricity mains. Many from the old school of electronics received electric shocks when fiddling around with open-valve-based radio sets and electronic equipment.

Since the valves themselves were made of glass, there was always the risk of them breaking for one reason or the other even though some were placed inside metallic protectors. All in all then, electronic equipment of the early days were heavy, fragile, and needed clean power. Certainly, not a combination of factors that would allow for portability and reliability! A typical valve-based radio, for example, would be approximately 20 inches wide, 15 inches high and 7 inches or so deep. A normal three-band-valve radio, common in most homes, could weigh from 5 to 6 kilograms.

As the complexity of the equipment increased, say transitioning from radios to television sets and onto computers, the size and weight of the equipment would increase dramatically. The British "Colossus" computer at the famous Bletchley Park of WWII fame had 1,500 vacuum tubes and associated circuitry. It occupied a large room and is reported to have weighed about a ton.

The ENIAC computer of the University of Pennsylvania (see chap. 6), had a staggering 18,000 vacuum tubes and weighed over 27,000 kilograms. It had a weight of 27 tons and needed 150 kilowatts of power to operate. Several engineers were required to change some 2,000 vacuum tubes each month. Now all this may have been sort of acceptable for number-churning, land-based, large computers of the time. But what about computers required on ships and aircraft?

By 1939, a new class of vacuum tubes and valves, called "miniature tubes" (Type B7G), were developed by RCA. With reduced sizes and reduced power requirements, it was now possible to make smaller consumer electronic items such as radios for home use as also some military equipment. RCA even came out with what they called the BP-10 "personal radio." This had a reduced size of roughly 9 × 3 × 4 inches and weighed a relatively lighter 2 kilograms. Yet, the batteries that were required to produce the high voltage would frequently run out.

At about the same time, Philips in Holland had also developed a very efficient miniature tube, the EF 50 and made their own machinery to manufacture this in large numbers. It is believed that in May 1940 with German troops massing on the border, a whole truckload of the machines along with drawings and raw material was loaded on to a truck, which sped away to put its cargo on a ferry to Britain the evening before the Germans came and captured the factory. The EF 50 was used in large numbers in the military equipment of the allies and is known as "the tube that helped to win the war."

The EF 50 miniature tube had by the 1950s become reliable enough to be used in computers. The very first major project to use them was Alan Turing's pilot ACE computer. But this was a laboratory-made computer. Possibly the very first commercial computer to use the new miniature vacuum tubes was the British-made Ferranti Mark 1, which used 1,600 of these vacuum tubes. The computer required only 25 kilowatts of power, somewhat modest for its time but still quite a lot.

As we have already noted, vacuum-tube-based electronic equipment had serious shortcomings. As far as the common user was concerned, all one would get would be a radio or music player in a fixed position at home or in the office. True, the development of the miniature valve enabled some reduction in size and the possibility of operating on batteries. Some car radios also now became feasible. Yet, the obvious shortcomings negated true portability and high reliability of the equipment. There was no way one could have music "on the go," so to say!

WWII also clearly brought out problems with vacuum-tube-based electronic equipment for the armed forces. The large size and high-power requirement were obvious enough issues. The fact that the vacuum tubes needed very frequent replacements, either due to "burn out" or just simple breaking of the fragile glass required a number of spare vacuum tubes to be always available in ready inventory. As far as computers were concerned, these problems multiplied manifold. A major technological breakthrough was once again desperately required, something literally more "solid" than the fragile and energy-hungry vacuum tubes.

Way back in 1833, the great Michael Faraday had noted that the conducting power of silver sulfide increased with an increase in temperature whereas the conducting power of metals would actually decrease with increasing temperature. Some 40 years later, in 1873, Willoughby Smith, a British engineer working for a cable manufacturer, noted another strange phenomenon. He noticed that the conductivity of selenium would increase with an exposure to light or what today is termed as the phenomenon of "photoconductivity."

In 1874, the Nobel Prize winner Karl Ferdinand Braun and Arthur Shuster (a scientist of German origin in England) working on galena and copper oxide, respectively, observed what today may be described as the process of current "rectification." William Adams and Richard Day, working further on selenium, found in 1876 that by the application of light to it, a small voltage could be generated—the "photovoltaic effect." Charles Fritts, the American inventor, would take this study further to produce an actual working "photocell," albeit with an efficiency of just about 1 percent.

Thus, before the end of the nineteenth century, "experimenters," if one may use the term, had noticed the phenomena of (1) negative temperature coefficient of resistance, (2) rectification, and (3) photovoltaic effect, in different materials. These observations seemed to indicate that there were some materials that had properties at variance from those of pure metals (conductors) as well as of pure insulators. The actual theoretical work behind these observations, however, had not yet been done but what was being noticed was the behavior of a class of material that came to be known as "semiconductors." Interestingly, the term "semiconducting" was first used by Alessandro Volta as early as 1782.

But the various observations and experiments on these materials with promise, the semiconductors, needed to be turned into something useful and practical as far as electronics was concerned. The first one off the mark was the Indian scientist, Jagadish Chandra Bose, about whom we have read in chapter 3 of this book.

Bose in his experiments needed to make a detector (diode) for radio waves. He settled on the use of galena (lead sulfide) crystals contacted by a metal whisker or ribbon for the purpose of signal detection and thus invented the crystal radio detector. Bose filed for a patent in 1901, which was formally received in 1904—the very first in the world for a "semiconductor device." Sadly, Bose being somewhat of an idealist did not set up any commercial entity to utilize his patents. He firmly believed that all inventions should be for the common good of mankind.

From 1902 through 1906, Greenleaf Pickard an engineer at AT&T, followed up on the pioneering work of Bose. He worked on many crystal materials to find the most efficient ones. Among these were silicon, molybdenum sulfide, and zincite. Pickard received in 1906 the first patent for a silicon-based device, a detector with a spring-loaded metal point touching the silicon crystal.

In order to commercialize his work on radio wave detectors, Pickard set up his own company in Boston, Wireless Speciality Apparatus Company in 1907, using the brand name of "Pericon" (*Per*fect *Pi*ckard *Con*tact). This company went on to produce crystal-based radio sets as also radio and other equipment for the military. Later, in 1922 it was acquired by RCA.[1]

There is then the rather interesting work done by a retired brigadier general of the US Army. Henry Harrison Chase Dunwoody was a former Army Signals officer and a graduate of West Point. After retirement, Gen. Dunwoody was hired by Lee de Forest in his company, the de Forest Wireless Telegraphy Company Inc. While employed there, in 1906, Dunwoody developed a crystal detector based on carborundum (silicon carbide) placed between two metallic steel sheets and received a patent for it. Marconi would later use these in his radio circuits. It may be noted that this invention may have helped save the de Forest Wireless Telegraph Company Inc. from losing a patent infringement case filed by the Canadian inventor Fessenden. Fessenden had claimed that de Forest had violated his patent by using the Fessenden electrolytic detector,[2] but the company could show that it was actually using Dunwoody's invention.

In the 1920s, a Soviet scientist by the name of Oleg Losev was also trying his hand at different materials for point-contact crystal detectors. After some experimentation, he settled on zincite (zinc oxide), to which he attached a steel point contact. Losev called his invention the "crystadine" negative resistance diode. This diode was then extensively used in Soviet-built radio receivers of that time. Losev would, however, become more famous for another discovery—the light emitting diode. We will read more about this later in the chapter. Tragically, Losev perished in WWII in the siege of Leningrad at the hands of the German armies.

In 1922, Lars Grondahl, an engineer working with Union Switch and Signal (a Westinghouse company) developed a copper–cupric oxide rectifier and received a patent for it. The device comprised a copper washer with cupric oxide formed on one side, which acted as the rectifying junction. By increasing the size of the copper washers and stacking more of them, larger currents could be handled. As a result, Grondahl's device could now be used in radio receivers as also be used as a rectifier for charging automotive batteries.

We have already noted Pickard's early work using silicon. In 1930, a scientist at Bell Labs, Russell Ohl, was tasked to work on developing an efficient crystal detector as a viable alternative to using a vacuum tube detector. Ohl studied and experimented with many materials, and much like Pickard, found that silicon clearly had the best potential. The problem, of course, was to find silicon of the right purity. Ohl then worked with metallurgists at Bell Labs to develop a good and efficient purification process. By 1940, reasonably high purity silicon was made available, but WWII put an interim halt to the program.

With Britain expecting an imminent invasion by the Germans in 1940, the foremost priority was to develop their own radar system (see chap. 5). With the use of the newly developed "cavity magnetron," it had become possible to use very high frequencies. But at such high frequencies, the use of vacuum tubes was found somewhat lacking in being able to detect signals bounced off flying aircraft. The solution lay in the use of solid-state point-contact detectors.

Two British companies, British Thompson Houston and General Electric Company (UK), were assigned the task of making silicon-based diode detectors. Unfortunately, neither company had the technology at that time of purifying silicon. They used commercially available polysilicon, which made the diodes quite unreliable. The assignment was promptly transferred back to the United States where Bell Labs developed the purification technology and manufacturing of the silicon detectors was transferred to their sister company, Western Electric within the AT&T group.

The great Russian scientist, Mendeleev, the father of the periodic table had predicted as early as 1869 the existence of an element that should lie between silicon and tin. Mendeleev termed this element as "ekasilicon." In 1886, Clemens Winkler, a German scientist from the Freiberg University of Mining and Technology, was asked to analyze a mineral from a nearby mine. Winkler first called this mineral "argyrodite," but on further experimentation he found that this was the ekasilicon predicted by Mendeleev and gave it the name of "germanium." It was, however, only in 1925 that Ernest Meritt of Cornell University made a working diode out of germanium.

By the 1940s, there were reasonably good solid-state alternatives to the vacuum-tube-based diodes and detectors. But a similar replacement for the main vacuum tubes used, for example, for amplification was nowhere yet on the horizon.

The "Transistor"

During WWII, the Bell Labs teams working on semiconductor-type material had been disbanded. Many of the scientists had gone off to work on defense-related projects at other laboratories and universities. However, after the war, in 1946, several of these scientists returned to Bell Labs and a new group constituted to work on semiconductors. This new team was headed by William Shockley and Stanley Morgan. The team members included Walter Brattain, John Bardeen, and a few more.

The team decided to concentrate on two materials only which had maximum promise—germanium and silicon.[3] The aim was to take forward from the earlier work done by Ohl to try to develop a sort of "signal processor" using a combination of material with different levels of impurities in them. By November 1947, Shockley's team had a working signal-processor device using germanium, albeit somewhat crude.

Before the end of the year, a better device using point contacts was developed. The then head of electronics research at Bell Labs, Dr. John Pierce was asked to give a name to this device. The name he came up with was the "transistor" (derived from *trans* resistance and var*istor*).[4] Pierce later became even more famous for designing the first communications satellite the "Telstar 1." This invention was announced to an astounded world only in June 1948, as it is believed that the Department of Defense wanted to keep this development completely secret for as long as possible.

The credit for the development of the germanium transistor was given only to Bardeen and Brattain. Shockley happened to be traveling at that time, and in his own words had the following to say, "The birth of the point-contact transistor was a magnificent Christmas present for the group as a whole. I shared the rejoicing. But my emotions were somewhat conflicted. My elation with the group's success was tempered by not being one of the inventors."[5]

Over the next few months, Shockley not only developed the full theory behind an even better transistor but also by 1950 he had actually fabricated one, termed as the "bipolar junction transistor," something that would completely revolutionize electronics. In 1956, the world applauded when Bardeen, Brattain, and Shockley were awarded the Nobel Prize for their work on transistors.

Of course, the transistor now needed to come out of the laboratory and into commercial production to be of real significance. Bell Labs licensed out the transistor technology to what they termed as "all responsible entities" at the incredibly low price of $25,000 each, upfront fee payment. In many cases, the royalties were adjusted against this upfront fee.[6] Western Electric quite naturally was one of the first licensees and so was Raytheon along with 28 others. It was Raytheon that produced the first commercially available transistor when at the end of 1948 they brought their CK 703 to the market. Western Electric opened their transistor manufacturing line in Allentown, Pennsylvania, only in 1951, giving Raytheon a free run of the market for some time. It was to be 1952 before another company, the Germanium Products Corporation (GPC), brought to the market their range of transistors made out of germanium. Unfortunately, within a short time with silicon rapidly overtaking germanium as the material of choice for transistors, GPC passed into oblivion.

Gordon Teal was one of the scientists working at Bell Labs and helped the Shockley team by developing the process of purifying germanium. In 1952, Teal left Bell Labs to join a company called Geophysical Services Inc. in Texas, a company then involved in oil field services. He then proceeded to set up a transistor manufacturing facility at the company after acquiring a license from Bell Labs. Shortly thereafter, the name of this company was changed to Texas Instruments (TI).

By 1954, TI had set up a full-fledged manufacturing facility not only for germanium transistors but for silicon transistors as well. Almost simultaneously, the company launched a subsidiary, Regency Electronics, to mass-produce radios using transistors produced by TI.

Meanwhile, the application of transistors in consumer electronic equipment had already started in 1952. Arguably, the first such product, a hearing aid, was produced by the Sonotone Corporation of New York. Raytheon and GPC by then had started to mass-produce transistors specifically for hearing aids. Later the same year, RCA demonstrated the very first experimental transistorized television receiver.

By 1953, the transistor radio had gained tremendously in popularity. A fledgling new Japanese company called Tokyo Tsushin Kogyo started the manufacture of really low-cost transistor radios, which they exported to the United States in large numbers. This company in 1955 changed its name to Sony Corporation. Also with new transistor radios in the market were GE, Raytheon, Zenith, and many others.

With the flurry of demand for transistors particularly of the silicon variety, many companies around the world started the manufacture of silicon transistors. William Shockley left Bell Labs in 1955 to form Shockley Semiconductor Laboratory as a division of Beckman Instruments. The company with premises located at Mountain View, California, had some extraordinarily talented engineers and scientists including Gordon Moore, Robert Noyce, Jean Hoerni, Jay Last, and others. Their quest was to produce the finest transistors in the business.

The Integrated Circuit: Fairchild, Texas Instruments, and National

Unfortunately, Shockley's management style was somewhat abrasive and tended to emphasize the research side of activities rather than focusing on making and selling products at a decent profit. The story has it that several of the good engineers and scientists asked Beckman Instruments to replace Shockley with a better manager. When their request was denied, this group (later dubbed "the gang of eight") led by Robert Noyce quit and in 1957 with the help of venture capital financing from the Fairchild Camera Company established Fairchild Semiconductors in a disused warehouse just down the road from Shockley Semiconductors.

Fairchild Semiconductors along with HP, Lockheed Missiles Division, with, of course, Stanford University, and the Xerox's PARC, all based in the San Francisco Bay Area, became the basic nucleus around which a very large number of high-technology industries and enterprises grew in that geographical area, earning it the name now etched in the world's collective minds—"Silicon Valley."

By 1958, Fairchild was producing the first of its range of transistors and the first batches were sold to IBM. Right from the start, their focus was on silicon-based devices. However, the technology levels of that time had limitations on the reliability of transistors produced, largely because of the transistor surfaces being unprotected from contamination and extraneous deposits. Jean Hoerni, a scientist of Swiss origin, later in 1958, developed at Fairchild a process that would make the silicon devices inherently reliable.

This process was termed as the "planar" process in which the interfaces (or junctions) on the semiconductor material were all on one plane and could be protected against contamination by a protective layer of silicon dioxide. Furthermore, several transistor elements could now be made on a single slice of silicon, and then physically separated by a process of cutting through the silicon. Each little piece of transistor element was then a small piece out of a bigger silicon disc or "wafer." Hence, the term "chip" came into being for the small transistor piece, as in "chip off a whole."

The planar process, in addition to enhancing the reliability of the transistor also enabled the use of thin-film metallic interconnections all on the surface. This meant that if the silicon surface was large enough, one could incorporate, for example, multiple transistors on it and have them interconnected by a process of vapor deposition of a metal. This procedure was then used effectively by Robert Noyce to make a silicon integrated circuit (IC) at Fairchild in 1959. In the years to follow, the company would become a world leader in many types of ICs.

Fairchild and some of the people that worked there would go on to spawn some of the greatest companies in the semiconductor and IC's business. These included names such as Intel, Advanced Micro Devices, National Semiconductors, Intersil, Very Large Scale Integration (VLSI) Technology, LSI Logic, Xilinx, Amelco (by Jean Hoerni), and others. Fairchild executives also went on to form some of the leading venture capital companies in the United States. These included, for example, names such as Sequoia Capital and Kleiner Perkins, Caulfield and Byers.

Yet, though Fairchild may have developed the concept of the planar IC, the credit for being the first to develop a working IC must go to Jack Kilby at TI. Kilby, in 1958, was a new employee at TI and was tasked to work on a program to develop modular circuits for the US Army. Kilby figured that rather than making the required circuitry using modules he could just as well incorporate the resistors and capacitors along with the transistor on the same piece of germanium. The external connections could then be provided by fine wires.

By September 1958, Kilby had demonstrated a working prototype of the world's first IC.[7] The world has not been the same since! Jack Kilby received the Nobel Prize for his invention in the year 2000. Following the development of the IC, the management at TI tasked Jack Kilby to produce an electronic gadget for more common use that would use multiples of his ICs than just the handfuls going into defense equipment. Shortly thereafter, Kilby designed the world's first electronic handheld calculator that would replace the clunky and large electromechanical desktop machines then in use.[8]

While Kilby, perhaps rightly, got the credit for developing a practical IC, it is only the very few, well versed in the technology, know that the concept of this great invention was first mooted in Britain as early as 1952. Geoffrey Dummer, then a scientist at the Telecommunications Research Establishment at Malvern, was convinced that the available transistor technology could be adapted to put multiple components on the same substrate. He clearly enunciated this concept at a technical conference held in Washington, DC, in 1952. This was the first public announcement of the concept of the IC.

In 1959, Dr. Bernard Rothlein left the Sperry Rand Corporation, to set up a semiconductor manufacturing company. He named it National Semiconductor Corporation (National). The company started operations in Danbury, Connecticut, and made a small range of speciality semiconductor products. The newly formed company was bedevilled by very low sales and by a patent infringement suit filed by Sperry Rand. In 1965, Peter Sprague, the son of the chairman of Sprague Electric Company, came to the help of National as a financier and became the chairman of its board of directors.

Peter Sprague moved the company to Santa Clara, California, in Silicon Valley. One of his first actions was to acquire a company called Molectro, which brought into the National fold four top former engineers from Fairchild including Robert Widlar, an expert in linear circuits. In 1967, National also managed to entice away five more top persons from Fairchild including one of their best managers, Charles Sprock, who was attracted by his new designation as president and by the allocation of stock in the company. Under this new management, the company really took off with sales reaching $40 million by 1970 and $365 million by 1976—this at a time when giants like GE and Westinghouse floundered in the highly price-competitive semiconductor business.[9]

With burgeoning competition from low-cost semiconductors now coming into the market from the Far East, National decided to get into the business of making electronic products including calculators, watches, games, and point of sale terminals. Incredibly, the company also tried to get into manufacturing digital computers to be sold in cooperation with Hitachi and a company called Itel.[10] This step was a disaster, as the ever-vigilant IBM was not going to brook such competition. This foray into manufacturing of computers by National was quickly terminated.

By 1981, National was in some trouble. Although its semiconductor activities were doing well, there was discontent within the company. Several mangers and engineers in particular the chief device designer, left the company. To get over these difficulties, National in an extremely bold and audacious move, bought out Fairchild at a staggering sum of around $120 million. But this move did not help either.

Drastic changes in top management were effected; a revised product mix was put in place. To buttress their IC business against competition National went on an acquisition spree taking over Mediamatics, Cyrix, ComCore Semiconductors, and others. Meanwhile, Fairchild was sold off to an investment company. But the sins of the past would keep catching up with the company. Eventually, in April 2011, National, one of the great semiconductor companies of our times, was sold off to TI.

Advanced Micro Devices (AMD)

In May 1969, another group of top people from Fairchild led by Jerry Sanders, then the international head of marketing left the company to start their own enterprise with a princely sum of $100,000. They named the company Advanced Micro Devices (AMD). The business model of the company was to reverse engineer popular semiconductors and ICs from established companies such as Fairchild and National, improve their specifications somewhat and become a second-source supplier to manufacturers of computers, instrumentation, and communication equipment.

Tasting early success, the company went public in 1972 and raised about $8 million. Before the end of the fifth year of operations, the product catalog offering was of a few hundred types of ICs. Even during the slowdown in the market in the mid-1970s, the company was averaging a healthy double-digit growth unlike some of its competitors. This robust performance attracted a $30 million cash for equity infusion by Siemens, which was desperately seeking a toehold in the United States. By 1978, the company had achieved the $100 million sales mark. The company was now on a roll!

By 1985, things started to get difficult for AMD. Manufacturers in the Far East, especially Japanese companies were now churning out competing products at much lower prices. To obtain newer products and technologies, AMD acquired a company called Monolithic Memories, but at $425 million worth of AMD shares, the purchase was perhaps overvalued. Shortly thereafter, Siemens divested its shareholding in the company. Despite these adversities, the company's progress stayed well on course and actually developed products that could compete with Intel, the market leader in microprocessors (see below).

In 1996, AMD splurged $857 million in stock to buy out a small chip-design company called NextGen, who had a great set of products, some designed by an engineer of Indian origin, Vinod Dham, who was widely acknowledged to be the father of the "Pentium" chip, one of the most famous ICs of all time, which he worked on at Intel. AMD was once again on a roll!

In 2006, AMD made what turned out to be a big mistake! It made a $5.4 billion purchase of a company called ATI Technologies, which was about 50 percent of AMD's own market capitalization at that time. The product program of ATI was good but the deal was extraordinarily overpriced. The gamble failed badly. As a result, AMD still plods along to a somewhat uncertain future despite a cash infusion by "Mubadla," an entity belonging to the Abu Dhabi Royal family, for 15 percent of the AMD stock.

Arguably, the biggest and the most famous of all the great semiconductor companies spawned by Fairchild was Intel Corporation. Two of the founders of Fairchild, Robert Noyce and Gordon Moore, teamed up with another Fairchild colleague, Andy Grove, to set up a company in 1968 called NM Electronics ("N" standing for Noyce and "M" for Moore) to manufacture large-scale ICs. Very quickly, they received venture capital financing, changed the name of the company to Intel (*Int*egrated *el*ectronics), and within a few months of starting, had produced an IC "memory" device that could potentially replace the magnetic-core memories then in use in computers. Intel then came out with improved semiconductor memory products at competitive prices validating their original concept that with the better speed and efficiency performance of ICs they would quite soon completely replace magnetic memories.

By 1971, the company had gone in for a public issue. The same year they had a somewhat lucky and fortuitous customer request. Busicom, a Japanese manufacturer of calculators requested Intel for a unique set of ICs for their planned new product range of calculators. This project was assigned by Intel to one of their really bright engineers, Ted Hoff, who came out with an absolutely novel design of incorporating over 2,000 transistors onto a single semiconductor chip measuring only 1/8 × 1/6 inches. This single chip, later dubbed the 4004 microprocessor, made history by being the world's very first computer processor on a single chip. It had the same computing power of the old ENIAC computer, which had approximately 18,000 vacuum tubes in it. Once again, history had been made![11]

In a matter of three years, Intel in 1974, brought to the market a whole "computer on a chip" called the 8080 microprocessor, which would rapidly become the industry standard. This product was sold at around $350 whereas standard computers of the time were selling for thousands of dollars. With IBM having approved and selected for use one of Intel's microprocessors to go into their PC, the company became the de facto market leader.

Over the next few years, Intel became the predominant semiconductor microprocessor and memory manufacturing company in the world. It introduced new products at regular intervals, most of them becoming instant successes. From the initial complement of 12 employees, by 1983, the company had about 15,000 employees and annual sales exceeding $1 billion. In 1989, Intel introduced the 80486 Microprocessor, which had over one million transistors on a single chip. With 50 times the speed of the pioneering 4004 microprocessor, this new device was almost a mainframe computer on a single chip.[12]

But the real blockbuster was yet to come. In 1993, Intel came out with the "Pentium chip" microprocessor, developed, as we have noted above, by Vinod Dham. This incredible product now had 3.1 million transistors on a single chip and was 1,500 times faster than the original 4004 microprocessor. Backed by an aggressive marketing campaign with the logo "Intel Inside," the Pentium rapidly became virtually a monopoly item for the computer industry. Annual sales of Intel in 1993 increased from $5.8 billion to a staggering $8.8 billion and net income in excess of $2 billion.

Every so often, Intel would introduce a new product. The new microprocessors seemed to follow a pattern in as much as the number of transistors they contained. In fact, this is precisely what Gordon Moore then at Fairchild prior to joining Intel had predicted and what has now become globally famous as "Moore's Law."

Simply stated, Moore's Law enunciates, "The number of transistors incorporated into a (microprocessor) chip will double every two years."[13] Devices with more than a billion transistors already exist. It may then be inferred that the actual final number can only be a function of the technological limits of the equipment to manufacture the devices. But with even this technology rapidly evolving, we can only wait and watch where the future takes us.

In July 2012, Intel paid $3.2 billion to acquire a 15 percent stake in the Dutch company ASML, a company that specializes and indeed dominates the market for the manufacture of some of the precision equipment needed for manufacture of advanced semiconductors. Intel also agreed to pay a further $1 billion for use in research and development by ASML.[14]

Semiconductor and IC manufacturing technology quite naturally spread to several countries of the world. Several non-US companies acquired licenses as well as established joint ventures. As the prices of semiconductors dropped rapidly, assembly and testing of many of these devices started to go offshore to low-cost labor countries that were also doling out tax and other incentives.

The first to set this trend was Fairchild, which set up the first offshore semiconductor operation in Hong Kong as early as 1962. Other companies followed suit later and moved assembly operations to low-cost geographies such as Singapore, Malaysia, Philippines, Thailand, and Taiwan. A representative timeline of such offshoring of assembly work by semiconductor companies is given in Appendix 5 of this book.

Many scholarly texts and articles have been written about the growth and dissemination of semiconductor technology in what used to be called the "Asian Tiger Economies." What started only as basic assembly grew from that point onward to inspection and testing of parameters. A few years later, several companies upgraded their overseas operations to full-device design and fabrication. Many independent local companies were spawned as subcontractors or contract manufacturers. Several companies, including US, Dutch, Swiss, and German ones started to produce device production and test equipment in the Asian Far East. A whole new technology paradigm was thus born in the region!

During the years of the cold war face off, the Soviets realized that they had been beaten in the high-tech race primarily because of the superiority of US-origin computer as well as chip technology. As a result, the Soviets and their COMECON friends "stole" or reverse engineered a lot of semiconductor technology as well as the production and test equipment used in the manufacture of these devices.

The Soviets had a very special unit specializing in just this kind of skulduggery. Many were given some kind of a diplomatic cover. It was the KGB's "Directorate T." They even managed to compromise an engineer of Argentine origin working with AMD, Bill Gaede, to spirit out the latest semiconductor secrets to the Soviets and East Germany via Cuba. There was even an allegation that Gaede, who subsequently went to work at Intel, stole the whole design and process of the "Pentium Chip" and passed this on to the Soviets.

But among all this international semiconductor activity, the achievements of originally a small start-up in Britain is worth noting, principally because today many of the so-called handheld electronic products have a direct linkage to this entity—ARM Holdings, as they are now called. The interesting thing is that they do not manufacture devices themselves. They only design and license out the technology and the intellectual property.

ARM, the Great Little "Chip" Company

The history of ARM starts in 1978, when a company called Acorn Computers was set up in Cambridge, England, by Chris Curry and Hermann Hauser to produce small home computers including the popular BBC- B—a computer popularized for school learning by the famous British broadcast house. When the development team at Acorn started to work on the next generation of computers targeted at principally the British market, they found that none of the available microprocessors quite measured up to their desired specifications. Furthermore, the prices that were being asked were way beyond their budget.

A decision was then taken to develop Acorn's own processor. It was felt that all the computational capabilities built into commercially available processors were not really required for Acorn's target market and hence their own development would proceed on something termed as "Reduced Instruction Set Computing," RISC for short. Design and development activities started in 1983 in total secrecy.

By 1985, the design part was finished and the first prototypes were fabricated for Acorn at VLSI Technology in the United States. The processor using only 25,000 transistors performed even better than expected. The component was dubbed the "ARM-1" processor, with ARM being the short form of "Acorn RISC Machine." This was the world's first commercial RISC processor. The following year, the processor was improved upon. Unfortunately for the company, by 1985, the whole PC and small computer business had drastically changed with the easy availability at reasonable prices of IBM and Apple branded computers running on standard off-the-shelf microprocessors.

Acorn now left without a competing computer model was in deep financial trouble. A savior was found in Olivetti, the Italian office machines company. Olivetti had taken over Acorn for its computer products but was pleasantly surprised when they found out that there was a great repository of processor technology within the company. Rather rapidly, an ARM-2 processor was developed and incorporated into a new computer offering called the "Archimedes," but this product did not evince much interest in the market with IBM and Apple being so ubiquitous.

It was, sort of, then "back to the drawing board." It was clear to the management of ARM that their computer selling days were pretty much now behind them. The future lay in improving on their RISC technology, positioning this for use as an embedded processor in higher-performance electronic equipment manufactured by other companies. The first supplies of the improved processor, ARM-3, designed by Acorn and manufactured by VLSI Technology met with great success. So much so that even Apple decided to become part of the "action."

In 1992, Acorn, VLSI Technology, and Apple joined together in establishing a new company named ARM Ltd. (ARM). The "Acorn RISC Machine" now was renamed as the "Advanced RISC Machine" and this is what gave the new company its name. The company's objectives were now to concentrate on device development for ultimate use by system developers at large. The company would also license its designs to producers of IC chips who could then sell these devices using their own brands.

ARM's great success was to be the designing of the processors for Apple's handheld organizer "Newton." Later in 1992, GEC Plessey in the United Kingdom signed on as a fabricating partner. In March 1993, Sharp of Japan entered into an agreement to manufacture and market ARM-designed products. The crowning glory was to then have TI sign on as a partner. There was no looking back now. Today ARM has over a thousand partners, over 30 billion processors sold, and some 16 million being sold each day. There is hardly any significant handheld device, including mobile and smart phones that do not use an ARM processor. A company that almost went bankrupt has now become a world leader.

Light Emitting Diodes (LED's)

In 1907, Henry Joseph Round, a researcher at the Marconi Laboratories in the United Kingdom, was tasked to work on silicon carbide (carborundum) based detectors using "cat's whisker" contacts for radio sets. In his experimentation, he observed a very strange phenomenon. On applying a voltage between two points on the material, he noticed a light yellow light being generated. On increasing the voltage, the light emitted would become brighter. In some cases, Round noticed other colors of light including green, orange, and blue usually at the negative terminal of the applied voltage. Round at that time had no rational explanations but today we know that what he had seen was the very first crude "Light Emitting Diode" (LED).

In 1922, a Russian researcher, Oleg Vladimirovich Losev, also working on crystal detectors independently observed what Round had seen in 1907—a green light emanating from a steel—silicon carbide diode when voltage was applied to it. Losev did, however, document his findings along with photographs of the light emitting region, and thus may formally be regarded as the inventor of the LED.

In 1955, Rubin Braunstein working at the RCA Laboratories found emission of light during his work on a material called gallium arsenide. No further development in this area is recorded until the late 1950s when the US Army Signals Corps Laboratory in Fort Monmouth made a few LEDs emitting orange color. In the 1960s, they followed up by making some that emitted green, red, and yellow colors. All were found terribly inefficient.[15]

In 1961, Robert Biard and Gary Pittman, then at TI, developed the first real functioning LED. This invention, however, came about by accident as the pair of scientists was actually trying to make a laser diode. Furthermore, this LED emitted only infrared light, which humans cannot see. TI, however, decided to commercialize these infrared LEDs, and which continue to be successfully used until now in fiber-optic communication systems.

It was only in 1962 that Nick Holonyack, an engineer at GE and formerly a student of John Bardeen, invented the first visible-light LED—one that emitted a red color. Holonyack would receive a patent for this invention. Holonyack would later also invent the laser diode as well as the light dimmer.

The first real big push for LEDs came from the chemicals giant, Monsanto. One of their engineers, M. George Craford, developed an LED that could emit yellow light and followed it up by developing a red light LED many times brighter than that made by Holonyack. Monsanto had worked on a partnership with HP whereby the latter would produce the LEDs using material supplied by Monsanto. Unfortunately, this deal did not go through. Monsanto then themselves took up the mass-scale production of these LEDs primarily for use as indicator lights and subsequently in digital watches, calculators, and clocks. Monsanto subsequently sold off the whole LED business to General Instruments.

The real success for LEDs was, however, made by Fairchild. By 1972, this company was able to utilize the planar semiconductor process that it had developed and used it for manufacturing really low-cost LEDs. Higher light output was achieved as well as greater reliability and power efficiency. Many companies around the world would later go on to apply the same fabrication process. Yet, the then existing LED technology was still not good enough for general-purpose lighting and illumination. For this, a very bright white or "warm white" light was required out of LEDs.

The breakthrough came as late as 1992. Shuji Nakamura, a Japanese scientist working with Nichia Corporation in Tokushima, that year developed the first practical blue-light LED using a nitride-based compound semiconductor. The blue light could then be converted into white or "warm white" by passing through a yellow phosphor coating. The world now had an energy saving replacement for the power-gulping incandescent bulb. Nakamura was awarded the "Millenium Technology Prize" in 2006.[16]

Organic LEDs (OLEDs)

But technology was not standing still. The electronic equipment business was desperately looking to replace power-gulping displays, especially the large and cumbersome picture tubes used in television sets. The demand was for much thinner, brighter, and flexible displays that could be sort of folded over. In 2009, LED displays for use in television sets had been introduced with some success, but the industry demanded something even better. The solution lay in what are called "organic light emitting diodes" or commonly called OLEDs.

Electroluminescent organic materials, of course, have been known for some years. The first-known research into these materials was done by A. Bernanose at the University of Nancy, in France, as early as the 1950s. In the 1960s, this research was carried forward by Dr. Martin Pope at New York University. But it would be 1965 before tangible results would be obtained for electroluminescence in organic material.

The first such development came out of the work done at the National Research Council of Canada by two scientists, W. Helfrich and W. G. Schneider, working on anthracene. The same year, scientists at Dow Chemical obtained similar results from a mixture of anthracene, tetracene, and graphite powder. Both groups were to obtain patents for their work. In 1975, Roger Partridge at United Kingdom's National Physical Laboratory would be the first, however, to develop a polymer LED.

In 1987, two researchers at Kodak, Ching W. Tang and Steven van Slyke, developed the OLED. In its basic form, the OLED is a conductive organic-material (as described above) layer deposited between two electrodes. This "sandwich" is then placed on a glass sheet or any other suitable transparent material. As current is applied, the organic material goes through a stage of "excitation," and when returning to its normal state, it emits a bright light.

A new company was set up in Ewing, New Jersey, in 1994, to commercialize Kodak's OLED technology. This company, Universal Display

Corporation (UDC) developed a business model much like that of ARM, whereby UDC carries out technology and product development and then licenses these out for actual manufacture by other companies. The company claimed to have Princeton University, University of Southern California, and the University of Michigan as its technology partners.[17]

In 1997, this company demonstrated the first OLED-based, flexible, flat-panel display. In 2001, UDC entered into an agreement with Sony for developing OLED television monitors. Today, the company has some 30 agreements with companies from many countries including famous names such as LG, Samsung, Seiko, Toyota, and so on.[18] In 2000, Motorola also licensed some of their own OLED technology to UDC in exchange for an equity position in UDC.

It was only in the late 1990s that commercial products using this technology would start to be available in the market. The first to market was the Japanese electronics company, Pioneer, which in 1997 brought out an OLED display for its range of car audio systems. By the 2000s, many companies were getting into the OLED business either for manufacturing the displays or for using them in their products. These included Sanyo (in association with Kodak), Toshiba, LG, NEC, Canon, and several others including Ritek in Taiwan.

Osram, the German manufacturer, used OLED technology in 2010 to launch the first "white color" lighting panel. Subsequently, Philips used this revolutionary new technology in its pathbreaking "Lumiblade" lighting panel, which is under 2 mm thick and can be embedded into all kinds of surfaces including windows, tabletops, photopanels, frames, and so on. Lumiblades are hugely energy efficient and the bonus is that there is no use of the poisonous mercury.

But in terms of displays for electronic equipment, perhaps no other company has put so much faith and effort into OLEDs as has Samsung. For some years now, Samsung has been using a version of OLEDs called the AMOLED in its displays, especially in their "Galaxy" range. The AMOLED display is an active "matrix" of OLEDs, which are activated by an array of thin film transistors. Such a display has many advantages over the passive OLED displays, especially in brightness, image quality, and reliability.

Because of the very large requirement of AMOLEDs for its own products, Samsung has its own manufacturing done by Samsung Display, which is the world's largest in this field. Recent competition to Samsung has arisen by LG setting up its own facility as well as from the Japanese companies, Sony, Panasonic, and Japan Display. As we write, news is coming in that even Apple, long a hold out against using AMOLED technology, has filed for a patent for a smart phone with a wraparound AMOLED display.

Liquid Crystal Displays (LCDs)

Most of us are familiar with the sort of grayish displays in a seven-segment pattern (looking like the numeral "8") in older versions of digital electronic watches and clocks. These displays are formed by a material known as "liquid crystals." Liquid crystals are substances that are neither full liquids nor full solids, something like a thick "sauce." When an electrical field is applied, the "crystals" in the thick solution align themselves into patterns forming displays. Such liquid crystal displays (LCDs) appeared to be a good solution in replacing battery-draining fluorescent and other displays.

Sometime in 1888, Friedrich Reinitzer, a scientist at the Charles University in Prague, was working on derivatives of carrots when he noticed that one particular compound turned into a cloudy liquid when heated. On further heating, it became transparent.[19] Since Reinitzer could not quite figure out this observation, he consulted another scientist, Otto Lehmann. In his experiments, Lehmann confirmed that while the material in its cloudy state was still fluid, much like normal crystals this material could refract light. For want of a better name, Lehmann named it as "flowing crystal" or "liquid crystal."

Over the years, research on such materials continued at various laboratories and universities. In 20 years, almost 200 such flowing-crystal-type materials were found to have similar characteristics but they remained largely as just laboratory findings until 1962. In that year, a researcher at RCA, Richard Williams, started to tinker around with a few of these materials to see if he could use them in making some kind of a display. He was indeed successful and filed for a patent in 1962.

A group under George Heilmeier was formed at RCA to see if some practical product could be made out of the work of Richard Williams. This group worked on this project for a few years, and with different materials and dyes. In 1967, they had a working prototype of a glass sandwich with a liquid crystal inside, which when a voltage was applied to it gave out as a display the familiar tuning-test pattern of television broadcasts quite common in those days. The RCA group had succeeded in making a working LCD. Over the next few months, Heilmeier's group had developed working prototypes of an electronic clock and an aircraft cockpit display.[20]

Heilmeier and his colleagues tried their very best to convince top management at RCA to set up the full-scale manufacturing of LCDs. They even got some external funding and produced LCD-based prototypes of advertising displays, fuel dispenser displays, and a glare-proof rear-view mirror for automobiles. But, RCA's top brass were not quite convinced about the commercial potential of LCDs.

Out of frustration, many of the RCA group working on LCDs left to join a new start-up called Optel. Heilmeier himself went on to join the service of the US government in a career switch. Some years later, when RCA noted the growing global business in LCDs, they tried to revive their program. But it was too late. They had let the golden goose go!

Hayakawa Electric Company of Japan, later to be called Sharp, had heard about the LCD work at RCA. They were looking for a lighter, low-power consuming display for their range of electronic calculators. In 1969, they approached RCA to make LCD-based calculator displays for them but were told that perhaps watch displays may be possible (for which RCA were trying to interest Timex) but displays for pocket calculators were just not possible. Sharp decided that they would make the LCDs themselves and a group was formed under Tomio Wada to do this. By 1973, they had perfected the display. Sharp used this LCD display in their range of pocket calculators and history was made.

Meanwhile, the group that had left RCA for Optel not only developed and manufactured small LCD displays but also succeeded in manufacturing the very first LCD digital watch in 1970. Unfortunately, Optel did not last too long in the business as the Japanese had come to dominate this segment, first with the calculators of Sharp and subsequently the watches from Seiko as well as other electronic items from Sony. Regrettably, RCA, as we have read earlier in this book, went into oblivion, purchased by GE—the LCD perhaps may have saved them their fate!

In 1985, Matsushita (Panasonic) created the world's first-ever color television with a LCD screen followed shortly in 1988, by Sharp who made the first LCD-based portable color television called the "Crystaltron." Almost all major television manufacturers were now switching over to LCD-based television sets. Homes, hotel rooms, pubs, and bars were all now flaunting the latest in televisions with large, LCD, flat screens, which could easily be mounted on walls, and be viewed reasonably well even from acute angles.

Since then, the use of LCDs in televisions has multiplied rapidly. By 2007, they had outstripped the conventional cathode-ray-tube-based television sets. By 2008, the numbers of television sets with LCDs had increased by 33 percent and by 2010 the number of sets sold with LCDs had reached a staggering total of about 190 million out of an annual 250 million sets.

Touch Screen Displays

Handheld devices, particularly the smart phones and "tablets," however, required something different. It was essential to eliminate the annoying keys, which in addition to taking up space, were also prone to miscuing or wrong entries due to slipped finger movements.

In chapter 7, we read how Steve Jobs of Apple was very keen to use some kind of a screen that itself could be used for entering commands instead of pressing individual mechanical keys. The answer lay in what is now called a "touch screen."

The history of the touch screen starts, as in the case of many other electronic items, in Britain. In 1965, E. A. Johnson, a scientist at the Royal Radar and Signals Establishment (formed by combining the Telecommunications Research Establishment with the Army Radar Establishment) at Malvern, was tasked to work on developing more modern and efficient radar displays for air-traffic control. Johnson incorporated a sensor with a small current flowing through it, into a glass panel. By touching the glass with a finger, a voltage change took place in the sensor. A controller could sense this voltage change and transmit it to the computer or other electronic devices.

Dr. Sam Hurst working at the University of Kentucky in 1971 further refined the concept of Johnson's touch screen sensor. This new invention was called the "Elograph" for which the university received a patent in 1973. Hurst left the University of Kentucky to incorporate his own company Elographics to commercialize the patent. In 1974, this company developed and produced the very first resistance touch-based transparent screen. This technology would later become the industry standard for many years to come, and be used in all kinds of appliances including mobile phones and tablets until the advent of the more efficient capacitive touch screen.

Sometime in 1977, Siemens wanted the development of a curved-glass "touch sensor." They financed Elographics to make this product for them. The successful development of this product resulted in the first use of the term "touch screen." In 1994, the name of the company changed to Elo Touch Systems more accurately reflecting what the company was making.

The 1980s brought more development work on touch screens. The first multitouch device was developed by a scientist of Indian origin, Nimish Mehta, in 1982 at the University of Toronto. In 1984, Myron

Krueger again from the University of Toronto developed a system, which could track hand movements. Bill Buxton another scientist at the University of Toronto perfected Mehta's multitouch device using a "capacitive touch sensor." The technology for touch screens was now good enough to be adapted into actual electronic products. In 1983, HP brought to the market the HP 150, the first PC using a touch screen. In 1984, Casio the Japanese company introduced the AT 550, a watch cum calculator with a touch screen. Atari, the electronics game company of yesteryear produced the first "point of sale" cash register with a touch screen in 1986.

In 1993, IBM together with Bell South came out with the rather short-lived "Simon Personal Communicator Phone"—a pager, phone, emailer, calendar, calculator, fax, all-in-one with a touch screen. The very same year, Apple would come out with its very first touch-screen-based product. This was the Newton PDA, which had a handwriting recognition feature associated with its touch screen. By 1996, Palm had launched its "Pilot" series with touch screens. The touch screen technology had by now really caught on.

The best, however, was yet to come. Wayne Westerman was working for his PhD on multitouch surfaces under Prof. John Elias at the University of Delaware in 1997. In 1998, the two jointly set up a company called Finger Works, which specialized in products using gesture recognition (originally designed for people with carpal tunnel medical issues) and multitouch-type sensors. Its principal products were a "gesture pad" and a multitouch keyboard. These products were so good that by 2005 the company, despite being small and without adequate financial resources, ended up being acquired by Apple—technology, people, and patents, all of it.

Westerman and Elias were also hired by Apple as senior engineers. The technology they had developed would subsequently be incorporated into Apple's products including Macbook, iPhone, iPad, and the iPod, which would take Apple to soaring heights. Unfortunately, the one feature that we are all familiar with on Apple products, and which the late Steve Jobs was so fond of demonstrating—the "pinch to zoom" feature—is not really an Apple but FingerWorks technology.

The US patent office very recently rejected a claim by Apple for a patent for this technology. Almost certainly, this was developed by Myron Krueger at the University of Toronto (please see above) way back in 1984. As we write, the battle of patents between Samsung (largely a user of AMOLED screens) and Apple still simmers. The rejection of Apple's claim for a patent for "pinch to zoom" could complicate things for them.

Appendix 1: Company Histories in Brief: The Pioneers of Telegraphy, Telephony, Wireless, Radio, and Television

Bell Telephone Company (The Bell System), US

The Bell Telephone Company was formally incorporated in 1877 by Gardiner Hubbard, the father-in-law of Alexander Graham Bell. Thomas Watson was put in charge of the technical functions of the company with Bell himself not being involved in the day-to-day operations of the company. Later that year, the company set up its first telephone exchange in New Haven, Connecticut. Within just a few years, the company had licensed the setting up of telephone exchanges in every major city in the United States. These franchises together with the parent company became known as "The Bell System" (with a bell as the symbol of the system). By 1881, the parent company had bought a controlling interest in the manufacturing company, Western Electric, which was promptly turned into the monopoly supplier of equipment to the Bell System.

A few years later, the American Telephone & Telegraph Company (AT&T) was set up as a subsidiary to establish the long-distance communications network. By 1885, AT&T had set up the first of these long-distance lines between New York and Philadelphia. The network then rapidly expanded to cover most of the United States as well as Canada and was run more or less as a monopoly by Theodore Vail, the then president of the company. There were in all 22 local operating companies, which later became popularly known as "Baby Bells." In 1899, AT&T acquired all the assets of the parent company.

Although the Bell System maintained fairly good service standards, the people started to tire of the monopoly. Hence, when Bell's patents ran out in the year 1894, many new manufacturing and independent telephone services companies sprang up in the United States, most in the smaller towns and rural communities of the country. By the turn of the century, the independent companies had a total number of subscribers just a little more than that of the Bell System.

The parent company then established the first West Coast to East Coast telephone network in 1915 and somehow revived its fortunes. Unfortunately, this was short lived as shortly thereafter the "big depression" hit the United States. When WWII started, most of the parent company's resources and that of Western Electric were diverted to the war effort, including even the development and manufacture of gun direction systems and later, radar systems.

In 1949, the US Justice Department started antitrust proceedings against the conglomerate. The case dragged on for many years and it was only in 1974 that the US government moved to break up the Bell System. By then, the Bell System had well over hundred million installed and operational phones in the United States. Finally, in 1983, Judge Harold Greene passed an order splitting up the Bell System into seven separate entities.[1]

British Telecom, UK

British Telecom is recognized as the world's oldest communications service provider. It was incorporated in the year 1846 as "The Electric Telegraph Company" initially as a domestic telegraphy services provider in Britain and within a decade had developed an international network.[2] It merged with the International Telegraph Company in 1855 to become the Electric and International Telegraph Company. This entity along with what remained of several smaller British service providers were amalgamated and nationalized into the British "General Post Office" (GPO).

With the beginning of telephony in Britain in the year 1878, services were provided by a few private companies, principally the National Telephone Company. In 1896, the GPO took over all the long-distance telephone operations from the National Telephone Company and in 1912 the GPO became the government's monopoly supplier of telegraphy and telephony services in Britain. After a series of governmental reforms resulting in the British Telecommunications Act, 1981, all of GPO's telecommunications-related activities were transferred to the autonomous corporation British Telecom or BT, as it became known. By 1984, BT was formally privatized.

Ericsson, Sweden

The company was founded by Lars Magnus Ericsson (earlier a trainee at Siemens & Halske) in Stockholm in the year 1876 and delivered its first products, telegraphy systems, to the state railways the same year. The following year, in 1878, the company initially started repairing telephone sets and telephony systems principally from the Bell Telephone Company and subsequently took up manufacturing of its own range of telephones. In 1907, the company went on to establish a telephone factory in Buffalo, New York. Then in 1809, a subsidiary was set up in Mexico, but the long civil war in the country had serious effects on the subsidiary.

By 1923, the parent company had developed advanced technology for telephone exchanges and installed its first five-hundred-lines exchange that year, and the company continued to prosper for the next seven years until a financier, Ivar Kreuger, took control of the company. This person, a somewhat unsavory character, indulged in all kinds of questionable dealings, and it was only after he committed suicide in 1932 that it was found that he had secretly sold off the company to a competitor, the American corporation, International Telephone & Telegraph Corporation (IT&T), another company with a somewhat colorful history. It took several years for the company to get back to some normality. By then, the Wallenberg family had bought out IT&T and brought in additional financing.

By 1956, Ericsson had developed the first "mobile phone," which weighed over 40 kilograms and was the size of a good-sized suitcase. In the year 1970, Ericsson started work on the electronic telephone exchange, which later became the "AXE System," proved to be a great winner, and also laid the foundation of Ericsson's later success in mobile telephony systems where it remains in a leading position worldwide.[3]

Magnetic Telegraph Company, US

After having successfully demonstrated his telegraphy system, Samuel Morse received financing from the US Congress in 1843 to set up a "demonstration line" between the cities of Baltimore and Washington, DC. He was able to conclusively show success on this line set up on telegraph poles. In order to be able to commercialize his invention and sell licenses, Morse hired the services of the retired postmaster general, Amos Kendall. In a short time, Kendall was able to put together a group of investors and in 1845 the Magnetic Telegraph Company was incorporated.

By the following year, this company had set up a telegraph line between Washington, DC, and New York as well as a few licenses given to other operators. Two of the licensees were the New York and Mississippi Valley Printing Company and the New York and Western Union Telegraph Company. These two companies merged in 1855 and also acquired some other smaller operators, to form the Western Union Company, which became the dominant telegraphy company in the United States.

In 1859, the Magnetic Telegraph Company was acquired by the American Telegraph Company, which in turn also merged into Western Union in the year 1866.

Marconi Wireless Telegraph Company and Marconi International Marine Company, UK

The Wireless Telegraph and Signal Company was started as a small factory in London in the year 1898. In 1900, the name was changed to the Marconi Wireless and Telegraph Company. The Marconi International Marine Company was also established later that year when Marconi's maritime rights were sold to it. With the rapid growth in business, the company was moved to a purpose-built factory building in Chelmsford in 1912. This was the very first purpose-built radio factory in the world, albeit principally for radio transmission systems.

In 1915, the company had made an aircraft telephone transmitter and by 1919 it had entered into a joint venture to establish the Marconi-Osram Valve Company to make radio valves. In the year 1923, it had developed the "otiphone," a hearing aid for those with hearing difficulties.

Meanwhile, the company continued production of its limited range of radio sets with the brand name of "Marconiphone" until 1923 when it sold its technology and rights to a newly formed company, "Marconiphone," which itself came to be acquired by the Gramaphone Company in 1929.

In 1928, the British government decided to amalgamate all the wireless telegraphy-operating companies, including that part of Marconi into a single entity called Cable and Wireless Ltd. Its remaining business then started work on television systems but by 1934 the television business was merged with those of the Electrical and Musical Industries (EMI), a company incorporated in 1931 to take over the operations of the Gramophone Company. By 1946, all of Marconi's interests, including in the Marconi International Maritime Company and the holdings of Cable and Wireless in Marconi, were acquired by the English Electric Company, which itself ultimately in 1968 became a part of the General Electric Company (GEC, earlier the General Electric Apparatus Company started by G. Byng and Hugo Hirst in 1886) of the United Kingdom.[4]

The Marconi International Marine Company, as we have seen in chapter 3, became globally famous for one of their employees, David Sarnoff, being on active duty, through the *Titanic* episode relaying messages to ships rushing for rescue operations. We will read more about David Sarnoff in the history of Radio Corporation of America (RCA) in Appendix 2 of this book.

Philco, US

The Helios Electric Company was started in 1892 to manufacture carbon-arc lamps. In 1906, it was renamed as the Philadelphia Storage Battery Company and began the manufacture of vehicle batteries. In 1919, the name of the company was shortened to Philco when it started the manufacture of small batteries and chargers meant for the radio industry. In 1926, Philco went into manufacturing of its own range of radios. By 1930, employing mass-manufacturing techniques, the company had become the largest radio manufacturer in the United States. The company also expanded its product line to car radios and other domestic appliances. In 1947, a range of televisions was added to the product offering.

Because of its very popular car radios, in 1961, the company was acquired by the Ford Motor Company, and the name changed to Philco Ford. Ford then decided to sell the consumer appliances part of the company to General Telephone & Electronics Corporation (GTE) which in turn, in 1981, sold off Philco to Philips of Netherlands (see below), which company for long had coveted the "Philips" trademark owned by Philco in the United States.[5]

Philips, Netherlands

The father and son team of Frederik and Gerard Philips set up the Philips Company in 1891, in the town of Eindhoven, Netherlands, to manufacture a range of electrical products including carbon-filament-based lamps. In the initial years, the company struggled, but with Gerard's younger brother, Anton, an engineer joining the company in 1895, matters began to improve substantially. By the 1920s, the company was manufacturing vacuum tubes and other components.

In 1927 and 1928, Philips successfully set up radio broadcasting stations including one that was named the "Happy Station" (in later years one of the favorites of this author), in Hilversum, Holland. By 1930, the company had also started manufacturing a range of radio sets. However, trouble came in a few years with the German invasion and capture of Holland. Anton Philips and several Philips family members managed to escape to the United States with a large part of the capital of the company and there established the North American Philips Company (owners of Magnavox and Sylvania).[6]

After WWII, the company headquarters were moved back to Eindhoven and research and manufacturing activities were restarted in right earnest. Philips would then go on to achieve some spectacular successes such as the world's first compact audiocassette tape, the first combination radio and cassette player, the first home video recorder, and many other innovative products including the ubiquitous electric razor, the "Philishave." However, with growing competition from cheaper and in many cases better performing products from Japanese and South Korean manufacturers, Philips's market share in the consumer electronics sector started to rapidly dwindle from the late 1990s. Ultimately, early in 2013, Philips finally saw the writing on the wall and sold off its consumer electronics business to the Japanese company Funai, with which it had a long-standing relationship.[7]

Siemens & Halske (Siemens), Germany

In 1846, a young artillery officer and a part-time inventor, only 30 years old at that time, decided that there was a great future in telegraphy. Werner von Siemens came out with a basic working model of a new and improved telegraphy system, but then got hold of an old associate of his from the "Physics Society," a mechanical engineer named George Halske, to do the actual construction of a full system. The two then decided to set up a company to commercialize their system and found a financier, Johan Georg Siemens, Werner's cousin (and father of the person who would become the cofounder of Deutsche Bank).

Siemens & Halske started business in October 1847. By 1848, the company had successfully set up a telegraphy system between Berlin and Frankfurt and with a growing business, Werner promptly resigned his army commission. By 1850, the company went international and set up offices in London and in St. Petersburg, Russia, and later in Vienna, Austria.

By 1866, Werner von Siemens came up with his most successful invention, that of the dynamo, which became the first step of the company into electrical engineering and products, a field in which it emerged as one of the world's most successful companies and remains a great multinational entity until today. In 1903, Siemens & Halske teamed up with another German company, AEG to establish what would become Telefunken (see below), one of the world's major manufacturers of radio and television equipment of its time.[8]

Telefunken, Germany

At the beginning of the twentieth century, two rival groups in Germany were working on radiotelegraphy technology. One of them was a group from the German company AEG and the other part of the Siemens & Halske group under the direction of Dr. Braun (who was later awarded a Nobel Prize). On the advice of then Kaiser Wilhelm II the activities of the two groups was merged into a new entity in 1903, to be called the "Gesselschaft fur drahtlose Telegraphie mbh." The first customers for the new entity were the German Army and Navy. In 1904, the name of the company was changed to Telefunken and in 1923 the company began the manufacture of radios for the mass market.

The company at that time was also making transmitters, receivers, and directional equipment, some of which were used in the original Graf Zeppelin. On October 31, 1928, Telefunken displayed its first working television system in Berlin. The company was then given the honors of arranging all the sound and broadcast systems for the famous 1936 Berlin Olympic Games.

In 1938, Telefunken had delivered the very first fully electronic television studio and in 1941 AEG bought out Siemens & Halske and now controlled 100 percent of the company. The post–WW II years were very difficult for Telefunken. While they did develop some excellent military radio technology, with the defeat of the Germans their markets more or less dried up. In 1960, the company managed a revival in the television business after developing the PAL color television system, which became the standard system in most parts of the world. By 1967, the entity was merged into AEG and became known as AEG Telefunken, however, due to poor management and also due to intense competition by the 1980s, the company was in a state of decline and divisions and product lines were closed or sold off. The state of affairs became really bad by 1985 by when Daimler Benz bought out what remained of AEG

Telefunken and all operations were discontinued except for preserving the brand name.[9]

Western Electric Company, US

Shortly after finally receiving the patent for the telephone, after all the controversies and legal wrangles, Alexander Graham Bell needed to establish manufacturing for the product. The Boston, Massachusetts, based firm of Charles Williams was selected for the purpose. This was the same company where Bell's assistant, Thomas Watson, had put in an apprenticeship period. Unfortunately, Williams's company was unable to manufacture adequate numbers of instruments of requisite quality, so Bell was forced to look out for other manufacturers.

Elisha Gray, the famous inventor and one of those with a claim to the invention of the telephone, had in 1872 bought into a company in Cleveland, Ohio, named Shawk & Barton, making electrical equipment, fire alarms, as well as typewriters for the Remington Company. This company was then renamed as the Western Electric Manufacturing Company and began manufacturing telegraph systems for the operator, Western Union Company. The company was also selected to manufacture telephone systems for Western Union, who themselves were claiming a patent for the invention of the telephone.

Western Union, however, realized that they would lose the telephone patent battle to Bell and hence stopped Western Electric from manufacturing the instruments. This left Western Electric without a substantial part of their business and overtures were made to Alexander Graham Bell's, the American Bell Telephone Company. Although reluctant at first, American Bell finally in 1881 agreed to buy a controlling interest in Western Electric not only to get good manufacturing capabilities but also to acquire the rights of some of the valuable patents held by Western Electric including the John Irwin and William Voelker patents (principally related to the transducer part of the telephone instrument), which had the potential of disrupting some of Bell's own patents.[10] Western Electric then went on to make telephone instruments exclusively for American Bell (renamed AT &T, in 1899).

Western Union Company, US

As we have noted in chapter 2 of this book, Western Union started its operations as the New York and Mississippi Valley Printing Company, a licensee of Samuel Morse's telegraphy technology. Meanwhile, Ezra

Cornell had set up a competitor, the New York and Western Union Telegraph Company. After a period of intense competition, Cornell proposed that it would be best for both companies if they could merge into one larger entity to provide services coast-to-coast across the United States. Thus, in 1855, the combined new entity became the Western Union Telegraph Company. It then went on to acquire several smaller telegraphy operators but ran into fierce competition from the Atlantic and Pacific Telegraph Company whose control was taken over by a financier, Jay Gould in 1875. In 1881, Gould went on to take over control of Western Union as well.[11]

The company then went on to introduce the first stock ticker (in 1869) and its famous money transfer business (in 1871) for which it became renowned. It opted out of a fledgling telephony business, based on Edison's technology after it realized that it would not succeed against the Bell Telephone Company on patent claims. In 1908, AT&T gained control of Western Union, and brought considerable strength to the company as it could now combine telegraphy and telephony services. However, by 1913, due to antitrust legislative pressures AT&T had to separate itself from Western Union. In 1945, Western Union merged with one of its principal rivals, the Postal Telegraph Company.

Western Union introduced several innovative services such as the first customer charge card (in 1914), the first "singing" telegram (in 1933), the first commercial inner-city microwave communications system (in 1943), the first commercial satellite in the United States (in 1974), and the first disposable prepaid phone card (in 1993).[12]

By the 1980s, due to financial problems, Western Union started to sell off or close down its telecommunication businesses to concentrate on its highly successful money transfer and financial services activities. By 1994, the company had become a part of First Data Corporation, a global leader in payments processing.

Westinghouse Electric & Manufacturing Company, US (Radio Operations)

Westinghouse was set up in 1886 by George Westinghouse, a most accomplished engineer. It was, among other manufacturing activities and companies of the group, already a well-known name in the manufacture of a range of electrical appliances. The company had been involved in radio-related work during WWI, but this activity increased substantially in 1920 after buying out the International Telegraph Company, the successor to radio pioneer Reginald Fessenden's National Electric

Signaling Company (NESCO). Shortly thereafter, Westinghouse established its own radio broadcasting station KDKA in Pittsburgh, after it received a commercial license and other stations were added later. It also set up its own radio manufacturing line.

However, Westinghouse realized that it was in no position to compete against the burgeoning fortunes of the RCA and in 1921 decided to sell off the International Telegraph Company and all its radio activities to RCA in exchange for RCA shares and become a collaborator with RCA.[13] The remaining part of the company bought out the renowned broadcaster Columbia Broadcasting System (CBS) in 1995 and changed the name of itself also to CBS Corporation in 1997.[14]

Appendix 2: Principal Entities Associated with Thomas Alva Edison and Their Timeline*

1869

Pope, Edison & Company (Electrical Engineers and General Telegraph Agency)
Bankers' and Brokers' Telegraph Company
Financial and Commercial Telegraph Company
Gold and Stock Reporting Telegraph Company

1870

American Printing Telegraph Company
Gold & Stock Telegraph Company (formed acquiring Financial and Commercial Telegraph Company and American Printing Telegraph Company—later to be a subsidiary of Western Union)
Automatic Telegraph Company (Edison becomes principal inventor)
American Telegraph Works

1871

Newark Telegraph Works (later renamed as Edison & Unger Company)

1873

Edison & Murray Company (telegraph manufacturing company)

1874

Domestic Telegraph Company, New York (for district telegraphy and fire/burglary systems)

1875

American Automatic Telegraph Company
Gilliland & Company (for manufacturing electrical/telegraphy instruments and electric pens)

1876

American Novelty Company (to make duplicating ink developed by
Edison)
Edison's Electric Pen and Duplicating Press Company

1877

American Speaking Telephone Company (combining Edison patents
to Gold & Stock Telegraph Company and Elisha Gray's telephone
patent to Harmonic Telegraph Company)

1878

Edison Electric Light Company
Edison Speaking Phonograph Company
Edison Phonograph Toy Manufacturing Company

1879

Edison Telephone Company of Europe
Edison Telephone Company of Glasgow
Edison Telephone Company of London
Societe du Telephone Edison, Paris
Menlo Park Manufacturing Company (for manufacture of Edison's
patented medicines)

1880

Edison Lamp Company
Societe General des Telephones
United Telephone Company Ltd. (merger between Edison Telephone
Co. Ltd. and Bell's, Telephone Company Ltd.)

1881

Edison Machine Works (for manufacture of dynamos and motors)
Electric Tube Company (merged into Edison Machine Works in 1885)
Edison Electric Light Company, Cuba and Puerto Rico
Edison Electric Light Company, Havana
Oriental Telephone Company Ltd.
Edison Company for Isolated Lighting
Edison Gower-Bell Company of Europe Ltd.

1882

Bergmann & Company (to manufacture electric fixtures, sockets, etc.)
Edison Electric Light Company Ltd., UK
Compagnie Continentale Edison

Edison's Indian and Colonial Electric Company Ltd.
Edison Spanish Colonial Light Company
Societe Electrique Edison, France
Societe Industrielle et Commerciale Edison, France (for manufacture
of lamps)

1883

Electric Railway Company of the United States
Societe Generale Italiana di Elettricita Sistema Edison, Italy
Societe d'Appareillage Electrique, Switzerland
Edison & Swan United Electric Light Company Ltd., London
Deutsche Edison Gesselschaft
Thomas A. Edison Central Station Construction Department (for
central power stations)
Argentine Edison Light Company

1884

A. B. Dick & Company (for manufacturing and sale of Edison's
mimeograph)

1886

Edison United Manufacturing Company
Australasian Electric Light Power & Storage Company Ltd.
Edison Electric Light Company of Europe
Sims-Edison Electric Torpedo Company

1887

Edison Phonograph Company
Edison Phonograph Toy Manufacturing Company
Allgemeine Elektrizitats Gessellschaft (AEG), Germany
Edison Wiring Company

1888

New England Phonograph Company
North American Phonograph Company
Edison Phonograph Works

1889

Edison's Phonograph Company, UK
Edison General Electric Company
Edison Manufacturing Company
Sprague Electric Railway and Motor Company

1890

Edison United Phonograph Company
Automatic Phonograph Exhibition Company

1892

General Electric Company (from merger of Edison General Electric and
Thomson-Houston Electric Company)
Edison Bell Phonograph Corporation Ltd., UK

1894

Kinetoscope Company
United States Phonograph Company

1896

Thomas A. Edison Inc. (original name National Phonograph Company)

1898

Edison Bell Consolidated Phonograph Company
Edisonia Ltd., UK

1901

Edison Storage Battery Company

1902

National Phonograph Company Ltd., UK (became Thomas A. Edison
Ltd., in 1912)

1903

Pike Adding Machine Company (acquired by Burroughs in 1909)

1904

Lansden Company (to make electric vehicles)

1908

Edison Business Phonograph Company

1910

General Film Company

1913

Edison Kinetophone Company
Edison Storage Battery Supply Company
Edison Storage Battery Garage Inc.
American Talking Picture Company

Appendix 3: Principal Entities Associated with Alexander Graham Bell and Their Timeline

1869   Western Electric Company is formed by Enos Barton and Elisha Gray

1872   Bell's School of Vocal Physiology set up

1877   Bell Telephone Company formed

1879   Bell Telephone Company merges with the New England Telephone Company to become National Bell Telephone Company

1880   National Bell Telephone Company becomes American Bell Telephone Company

1880   Bell receives Volta prize from the French government and sets up Volta Laboratory

1880   Bell Telephone Company of Canada incorporated

1881   American Bell Telephone Company does a reverse merger with its supplier Western Electric Company

1883   Bell starts *Science* magazine

1885   American Telephone and Telegraph Company (AT&T) incorporated as a subsidiary of American Bell Telephone Company to operate the long-distance telephone network

1887   Bell becomes president of the "National Geographic Society"

1887   Volta Laboratory transfers the Phonograph patents to the American Gramophone Company, later to evolve into Columbia Records

1895   Bell Telephone Company of Canada spins off its manufacturing to the Northern Electric and Manufacturing Company Ltd.

1899   AT&T becomes the mother company of the Bell System including the various operators later to be called "Baby Bells"

1907   Western Electric Research Department created

1925 AT&T's Engineering Department merged with Western Electric Research Department to form Bell Telephone Laboratories Inc.

1925 International Western Electric Company (a subsidiary of Western Electric) sold to International Telephone & Telegraph Corporation (IT&T)

1939 AT&T controls 80 percent of all telephone connections, 98 percent of all long-distance phone lines, and 90 percent of all telephone equipment sold in the United States

1949 After an antitrust ruling, AT&T sells its stake in Northern Electric to Bell Canada

1964 Bell Canada acquires Northern Electric

1976 Northern Electric renamed as Northern Telecom

1984 After the demerger, AT&T becomes AT&T Technologies

1986 IT&T sells its European telecommunications business to Companie General d'Electricite. Name subsequently changed to Alcatel Alsthom.

1996 AT&T Technologies spins off its manufacturing into Lucent Technologies. Most of Bell Laboratories work transferred to Lucent.

1998 Northern Telecom becomes Nortel Networks

1998 Alcatel Alsthom's name changed to Alcatel

2000 Business Communications division of AT&T Technologies spun off as "Avaya"

2006 Alcatel and Lucent merge to form Alcatel-Lucent

2009 Nortel Networks files for bankruptcy. Ericsson picks up most of its telecom assets.

Appendix 4: Principal Entities Associated with Guglielmo Marconi and Their Timeline*

1804   *Elliott Instruments Company founded*
1885   *Ferranti is founded*
1889   *General Electric Company (GEC) of UK is founded*
1896   *British Thomson-Houston Company (BTHC)*
1899   *British Westinghouse Electric & Manufacturing Company (BWEM)*
1897   Wireless Telegraph & Trading Signal Company (WTTSC)
1900   Renamed: Marconi Wireless & Telegraph Company (MWTC)
1900   The Marconi International Marine Company
1901   Marconi Wireless Telegraph Training College
1902   Marconi Wireless Telegraph Company of America
1904   Marconi Marine Company
1905   Canadian Marconi Company
1908   Russian Company of Wireless Telegraphs and Telephones
1917   *Plessey is founded*
1918   *English Electric Company Ltd. is incorporated*
1919   *"BWEM" name changed to Metropolitan Vickers Electrical Company Ltd. (MVEC)*
1919   Marconi Osram Valve Company (MOVC)
1919   *Radio Corporation of America (RCA) is incorporated*
1920   RCA acquires Marconi Wireless Telegraph Company of America
1920   *RCA (UK) established*
1922   British Broadcasting Corporation (BBC)
1924   Unione Radiofonica Italiana
1929   Cable & Wireless Ltd. (amalgamation between MWTC and Eastern Telegraph Group)
1929   *Associated Electrical Industries (combines MVEC and BTHC)*
1929   MOVC sold to Gramophone Company (later EMI)

1936 Marconi Ecko Instruments
1939 Marconi Research Laboratory
1941 Marconi Ecko Instruments becomes "Marconi Instruments"
1963 MWTC name changed to Marconi Company
1948 English Electric acquires the Marconi Company
1961 *GEC acquires Radio & Allied Industries*
1967 Eddystone Radio (after taking over Stratton & Company)
1967 *GEC acquires Associated Electrical Industries*
1968 *GEC merged with English Electric Company (incorporating The Marconi Company)*
1968 Marconi Space & Defence Systems (MSDS)
1968 Marconi Elliott Space & Defence Systems (name changed from MSDS)
1968 Marconi Underwater Systems Ltd.
1969 Marconi Elliott Avionics Systems Ltd.
1969 Marconi Elliott Computer Systems Ltd.
1969 Marconi Communications Systems Ltd. (MCS)
1987 Marconi Company becomes GEC-Marconi
1988 *RCA (UK) becomes SERCO*
1988 *GEC and Siemens acquire Plessey to form GP Telecommunications*
1990 *GEC acquires Ferranti Electronics*
1998 GEC Marconi Communications—GMC (formed from amalgamation of GPT's global entities)
1998 Marconi Electronics Systems Ltd. (renamed from GEC Defence Systems)
1999 Marconi plc (from GEC's non defense business including Ferranti)
1999 Marconi Electronics Systems and British Aerospace merge to form BAE
1999 Marconi Communications becomes subsidiary of Marconi plc
1999 Marconi Medical Systems (out of GEC-Picker)
2000 Marconi plc becomes Marconi Corporation plc
2000 Marconi Mobile Ltd. (name changed from GMC)
2006 *Ericsson acquires Marconi Corporation plc*

Appendix 5: Timeline*—Offshoring of Semiconductor Assembly

Year of offshoring	Name of company	Where offshored
1962	Fairchild Semiconductor	Hong Kong
1965	Signetics	South Korea
1966	General Instruments	Taiwan
1967	Radio Corporation of America	Taiwan
1967	Texas Instruments	Singapore
1972	Texas Instruments	Netherlands Antilles
1972	Intel Corporation	Malaysia
1972	National Semiconductor Corporation	Malaysia
1972	Hewlett & Packard	Malaysia
1972	Texas Instruments	Malaysia
1972	Advanced Micro Devices	Malaysia
1973	National Semiconductor Corporation	Thailand
1973	Motorola	Malaysia
1973	Mostek	Malaysia
1973	Hitachi	Malaysia
1974	National Semiconductor Corporation	Indonesia
1974	Intel Corporation	Philippines
1974	Radio Corporation of America	Malaysia
1974	Harris Semiconductor	Malaysia
1974	NEC	Malaysia
1974	Toshiba	Malaysia
1974	Siemens	Malaysia
1977	Intel Corporation	Barbados
1979	Intersil	India
1979	General Instruments	Malaysia
1985	American Telephone and Telegraph	Thailand

Notes

Preface

1. "Modern Families: Chips off the Old Block," *The Economist*, January 12, 2013.

1 Introduction

1. Rev. C. W. King, *The Natural History of Gems or Decorative Stones*, Cambridge, UK, 1867.
2. Michael R. Collings, *Gemlore: An Introduction to Precious and Semi Precious Stones*, Rockville, MD: Borgo Press, 2009, ISBN 1434457028.
3. J. L. Heilbron, *Electricity in the 17th and 18th Centuries: A Study of Early Modern Physics*, New York: University of California Press, 1979.
4. William Gilbert, *De Magnete*, English translation by Paul Fleury Mottelay, New York: Dover Publications, 2009, ISBN 0–486–26761-X.
5. Mike Brand, Sharon Neaves, and Emily Smith, *"Lodestone": Museum of Electricity and Magnetism*, Tallahassee, FL: Mag Lab U. US National High Magnetic Field Laboratory, 1995.
6. Geno Jezek, "History of Magnets," Montana, 2006–2013, www.how magnetswork.com/history.html.
7. H. H. Ricker III, "The Discovery of the Magnetic Compass and Its Use in Navigation," *The General Science Journal*, July 25, 2005, www.gsjournal. net/old/science/ricker4.
8. Paul Fleury Mottelay, *The Bibliographic History of Electricity and Magnetism*, London: Charles Griffin & Co., 1922.
9. John B. Carlson, "Lodestone Compass": Chinese or Olmmec Primacy? Multidisciplinary Analysis of an Olmec Hematite Artifact from San Lorenzo, Veracruz, Mexico," *Science* 189, no. 4205 (1975): 753–60.
10. Maharishi Bharadwaja, *Vaimanika Shastra*, English translation by G. R. Josyer, 1973, Baroda, India: Rajkiya Sanskrit Library, 1944. David Hatcher Childress, *Vimana Aircraft of Ancient India and Atlantis*, Kempton, IL: Adventures Unlimited Press, 1991.
11. Dr. V. Raghavan, *Yantras or Mechanical Contrivances in Ancient India*, Bangalore: The Indian Institute of Culture, 1956.

12. Arran Frood, "Riddle of Baghdad's Batteries," BBC Website, February, 27, 2003.
13. National High Magnetic Field Laboratory, Florida State University, www .magnet.fsu.edu/education/tutorials/timeline/600bc-1599.html.
14. Carl Sagan, "Fine Art of Baloney Detection," in *The Demon Haunted World: Science as a Candle in the Dark*, New York : Ballantine Books, 1997, ISBN 0–345–40946–9.
15. Joseph Stewart, "Intermediate Electromagnetic Theory," Singapore: World Scientific Publications Co. Pte. Ltd., 2001.
16. From IEEE Global History Network, www.ieeeghn.org/wiki/index.php /Capacitors.
17. Luigi Galvani, *De Viribus Electricitatis in Motu Musculari*, 1792.
18. Notes and records of the Royal Society.
19. At the time of writing, the first-known recording made in June 1878 by Edison has just been unveiled. It is that of a man reciting the nursery rhymes "Mary Had a Little Lamb" and "Old Mother Hubbard," recorded on a sheet of tin foil, placed on the cylinder of Edison's phonograph.
20. "Hertz Biography," Institute of Chemistry, Hebrew University of Jerusalem. Heinrich Hertz; IEEE Global History Network, ghn.ieee.org/wiki6/index .php/Heinrich_Hertz_(1857–1894).
21. E. A. Davis and I. J. Falconer, *J. J. Thomson and the Discovery of the Electron*, UK & Boca Raton, FL: CRC Press, 1997.
22. Sir J. J. Thomson got the Nobel Prize in 1906. Extraordinarily, seven of his research assistants also got the Nobel Prize and amazingly so did his son, Sir George Paget Thomson, in 1937.
23. James A. Hijiya, *Lee de Forest and the Fatherhood of Radio*, Bethlehem, PA: Lehigh University Press, 1992, ISBN 978–0–934223–23–2.
24. Ministry of Foreign Affairs, *Virtual Finland*, Helsinki, Finland: Ministry of Foreign Affairs, http://virtual.finland.fi/netcomm/news/showarticle. asp?intNWSAID=25818.
25. Antti V. Raisanen and Arto Lehto, *Radio Engineering for Wireless Communications and Sensor Applications*, Boston, MA: Artech House Mobile Communications Series, 2003.
26. Irving Langmuir, IEEE Global History Network, www.ieeeghn.org/wiki/index .php/Irving_Langmuir.

2 The Early Years: Telegraphy and Telephony

1. Prof. J.M. Dilhac, "The Telegraph of Claude Chappe: An Optical Telecommunication Network for the XVIIIth Century," IEEE Global History Network, www.ieeeghn.org.
2. From the IEEE Global History Network, www.ieeeghn.org/wiki/index .php/Milestones:Schillings_Pioneering_Contribution_to_Practical _Telegraphy,_1828–1837.
3. The Samuel Morse website, www.samuelmorse.net/morses-telegraph/electrical -telegraph.

4. Steven Roberts, "Distant Writing: A History of the Telegraph Companies in Britain between 1838 and 1868," http://distantwriting.co.uk.
5. Ibid.
6. Prof. Angel Lazano, "The Hall of Innovation," Department of Information and Communication Technologies, Universitat Pompeu Fabra, www.dtic.upf.edu.
7. Brian Bowers, "Cooke and Wheatstone, and Morse—a Comparative View," Conference on the History of Electronics, IEEE History Center, 2001.
8. "Siemens History," www.siemens.com/history/en.
9. "Our History," Cable & Wireless Communications, www.cwc.com/past-present/our-history.html.
10. Rory Carroll, "Bell Did Not Invent the Telephone—US Rules," *The Guardian*, June 17, 2002.
11. "Elisha Gray—the race to Patent the Telephone," www.inventors.about.com/od/gstartinventors/a/Elisha_Gray.htm.
12. "Elisha Gray," Oberlin College Archives, www.oberlin.edu//external/EOG/OYTT-images/ElishaGray.html.
13. Sylvanus P. Thompson, *Phillipp Reis: Inventor of the Telephone*, London: E & F. N. Spon, 1883.
14. "Alexander Graham Bell," Wikipedia, the Free Encyclopedia, en.wikipedia.org/wiki/Alexander_Graham_Bell.
15. "'Imagining the Internet': A History and Forecast," Elon University School of Communications, http://www.elon.edu/e-web/predictions/150/1870.xhtml.
16. "Obama's Whopper about Rutherford B. Hayes and the Telephone," *The Washington Post*, March 16, 2012, www.washingtonpost.com.
17. "The History of Ericsson," www.ericssonhistory.com.
18. "UK Telephone History [from the British Telecom Archives]," Bobs Telephone File, www.britishtelephones.com/histuk.htm.
19. "Emile Berliner: The History of the Gramophone," http://inventors.about.com/od/gstartinventions/a/gramophone.htm.
20. "Alexander Graham Bell, Biography," www.biography.com/people/alexander-graham-bell-9205497.

3 Wireless and Radio

1. James Clerk Maxwell, "A Dynamical Theory of the Electro–Magnetic Field," Philosophical Transactions of the Royal Society of London, 1865.
2. "Etheric Force," IEEE Global History Network, www.ieeeghn.org/wiki/index.php/Etheric_Force.
3. "Oliver Lodge," IEEE Global History Newtork, www.ieeeghn.org/wiki/index.php/Oliver_Lodge.
4. James P. Rybak, "Oliver Lodge: *Almost* the Father of Radio," Grand Junction, CO: Mesa State College, www.antiquewireless.org/otb/lodge1102.htm.
5. Gerald Beals, "Major Inventions and Events in the Life of Thomas Alva Edison," www.thomasedison.com/Inventions.htm.

6. H. Hertz, *Electric Waves: Being Researches on the Propagation of Electric Action with Finite Velocity through Space*, Dover Publications, 1893.

7. "IMS 2012 International Microwaves Symposium, Famous Quotes," http://ims2012.mtt.org/en/node/188.

8. Father Landell Demoura, Brazilian Patent 3279, www.landelldemoura.qsl.br.

9. "Tesla Biography," Tesla Memorial Society of New York, www.teslasociety.com/biography.htm.

10. Ibid.

11. "Blackberry Predicted a Century Ago by Pioneering Physicist Nikola Tesla," *The Telegraph*, May 3, 2010, www.telegraph.co.uk/technology/blackberry/7674280.

12. Letter in the archives of Rabindra Bhavan, Visva Bharati University, Santiniketan, West Bengal, India.

13. Varun Aggarwal, "Jagadish Chandra Bose: The Real Inventor of Marconi's Wireless Receiver," Div. of Electronics and Comm. Engineering, NSIT, Delhi, India, http://web.mit.edu/varun_ag/www/bose_real_inventor.pdf.

14. "Russian Radio Pioneer Popov Honoured by ITU," November 2009, www.itu.int/net/itunews/issues/2009/09/57.aspx.

15. Ibid.

16. "Guglielmo Marconi, Wireless Communication," http://web.mit.edu/invent/iow/marconi.html.

17. "Tesla: Master of Lightning," Public Broadcasting System broadcast, April 2004, www.pbs.org/tesla.

18. Probir K. Bondyopadhyay, "Under the Glare of a Thousand Suns—the Pioneering Works of Sir J. C. Bose," *Proc. IEEE* 86, no. 1 (January 1998): 218–85.

19. "From On-Air to Disrepair: Battle to Save First Ever Wireless Factory from Where Marconi Broadcast to the World 100 Years Ago," *Daily Mail*, July18, 2012.

20. "Telefunken Company History, 1903–1922," www.telefunken.com/company/history/1903–1922.

21. John S. Belrose, "Fessenden and the Early History of Radio Science," *The Radioscientist* 5, no. 3 (September 1994): 94–110.

22. "Famous Last Words," Randy Winchester's Place, http://web.mit.edu/randy/www/words.html.

23. Gordon Grebs and Mike Adams, *Charles Herrold, Inventor of Radio Broadcasting*, Jefferson, NC: Mcfarland & Co., August 15, 2003, ISBN-10:0786416904.

24. "Crystal Radios: The First Wireless," PV Scientific Instruments, www.arcsandsparks.com/aboutcrystalradios.html.

25. Ralph Williams, "Atwater Kent: The Man and His Radios," www.atwaterkentradio.com/atwater.htm.

26. "The History of Powel Crosley's Crosley Radios," www.crosleyradio.com/about.html.

27. "The Pye Museum of Pye Telecom Products and Company History," www.pyetelecomhistory.org.

28. "History of the Radio Manufacturer Amalgamated Wireless Australia Ltd. (AWA)," www.radiomuseum.org/dsp_hersteller_detail.cfm?company_id=7394.

29. "First Car Radios—History and Development of Early Car Radios," www
.radiomuseum.org/forum/first_car_radios_history_and_development_of
_early_car_radios.html.
30. Ibid.
31. "Motorola's History and Knowledge," www.motorola.com/us/Motorola
-History/Corporate-Motorola-History.html.
32. "Edwin H. Armstrong, Biography," IEEE Global History Network, www
.ieeeghn.org/wiki/index.php/Edwin_H._Armstrong.
33. "Edwin Armstrong, the Creator of FM radio," The First Electronic Church of
America http://fecha.org/armstrong.htm.
34. Don Pies, "Regency TR-1 Transistor Radio History," Santa Barbara, CA,
www.regencytrl.com.
35. "About Sony, Sony Corporate History," www.sony-europe.com/artice/id
/1178278971500.

4   Television

*The term "television" is believed to have been first used by a Russian scientist,
Constantin Perskyi, at the Paris Fair, in the year 1900.

1. "Giovanni Caselli," www.telephonecollecting.org/caselli.htm.
2. George R. Carey Television Archive, 1878–1903, Manuscript Archive,
Michael Brown Rare Books, http://mbamericana.com/george-r-carey
-television-archive-1878–1903.
3. Maurice LeBlanc, "Etude sur la transmission electrique des impressions
lumineuses," *La Lumiere Electrique*, Paris, France, 1er Decembre, 1880.
4. Alexander Graham Bell, "On the Production and Reproduction of Sound
by Light—the Photophone," *Proceedings, American Association for the
Advancement of Science* 29 (Oct. 1880): 115–136.
5. Mary Bellis, "Television History—Paul Nipkow," http://inventors.about
.com/od/germaninventors/a/Nipkow.htm.
6. Christabel Donatienne Ruby, ed., *Eugen Goldstein*, Port Louis, Mauritius
and Dusseldorf, Germany: Fidel, 2011, ISBN 6138167402.
7. "Boris Lvovich Rosing (Russian) (1869–1933)," Baird Television, www.bairdtele-
vision.com/rosing.html; "Celebrities: Rosing, Boris L. (Russian Inventor of
Television)," http://persona.rin.ru/eng/view/f/0/34302/rosing-boris-l.
8. Hal Landen, adapted, "The Birth of Television," www.videouniversity.
com/articles/the-birth-of-television.
9. Ibid.
10. "Vladimir K. Zworykin (1889–1982): Electronic Television," February 2000,
http://web.mit.edu/invent/iow/zworykin.html.
11. Ibid.
12. "Charles Francis Jenkins (American) (1867–1934)," Baird Television, www
.bairdtelevision.com/jenkins.html.
13. "John Logie Baird: A Life," www.bairdtelevision.com.

14. "'The Televisor'—Successful Test of New Apparatus," *The Times*, London, Thursday, January 28, 1926, page 9.
15. "Next We'll See to Paris, 1927—Baird Television," www.bairdtelevision.com /1927.html.
16. "The DuMont Telecruiser: A Little DuMont Background," http://chalk hillmedia.org/telecruiser/DuMontHistory.htm.
17. "British Industrial History: Cossor Radio and Television of London," www .gracesguide.co.uk/Cossor_Radio_and_Television.
18. Mary Bellis, "Color Television History: The Development of Color Television," inventors.about.com/library/inventors/blcolortelevision.htm.
19. Mary Bellis, ed., "Cable Television History," inventors.about.com/library /inventors/blcabletelevision.htm.
20. Kevin Custer, "Jerrold Electronics," www.catvtech.com/miltonshapp.html.
21. "Tech-FAQ: The History of Satellite Television," www.tech-faq.com/history -of-satellite-television.html.
22. Christopher Klein, "The Birth of Satellite TV, 50 Years Ago," July 23, 2012, www.history.com/news/the-birth-of-satellite-tv-50-years-ago.
23. "Television History—The First 75 years: 1946–49, Garod—USA," www .tvhistory.tv/1946–49-GAROD.htm.
24. Ibid.
25. "Obituary: Eugene Polley," *The Times*, London, May 24, 2012.
26. Ibid.

5 World War II: Radar, Sonar, Cryptography, and Beyond

1. "BBC—History: The Battle of Britain," www.bbc.co.uk/history/battle _of_britain.
2. "Christian Huelsmeyer, the Inventor," Radar World, Martin Hollman, 2007, www.radarworld.org/huelsmeyer.
3. Major Gregory C. Clarke, "Deflating British radar myths of World War II," Air Command and Staff College, Montgomery, Alabama, March 1997, www.radarpages.co.uk/download/AUACSC0609F97–3.pdf.
4. Ibid.
5. Harry von Kruge, *GEMA—Birthplace of German Radar and Sonar*, Boca Raton, FL: CRC Press, 2002.
6. Martin Hollman, "Radar Development in Germany," Radar World, 2007, www.radarworld.org/germany.html.
7. Michael Portillo, "The Junkers of Woodbridge Airfield," www.bbc.co.uk /podcasts/series/twftr.
8. Anthony Sampson, *The Sovereign State: The Secret History of IT&T*, London: Hodder and Stoughton, 1973.
9. Rob Arndt, "GEMA Kristall Flugzeug, DFS-468, Schwarz Diamant (1945)," http://discaircraft.greyfalcon.us/Gema.htm.
10. John E. Gorham, "Electron Tubes in World War II," Proceedings of the IRE, Waves and Electrons Section, 1947.

11. J. T. Randall and H. A. H. Boot, "Historical Notes on the Cavity Magnetron," *IEEE Transactions on Electronic Devices* 23, no. 7 (1976): 724–729.
12. "Raytheon Company History," http://www.fundinguniverse.com/company -histories/Raytheon-company-history.
13. Dick Barrett, "Radar Personalities: Sir Robert Watson-Watt," December 18, 2000, www.radarpages.co.uk/people/watson-watt/watson-watt.htm.
14. Ibid.
15. John Whiteclay Chambers II, ed., *"SONAR": The Oxford Companion to American Military History*, Oxford, UK: Oxford University Press, 1999, ISBN 0–19–507198–0.
16. Angela D'Amico and Richard Pittenger, "A Brief History of Active Sonar," *Aquatic Mammals* 35, no. 4 (2009): 426–434.
17. Lt. John Howard, "Fixed Sonar Systems—The History and Future of the Underwater Silent Sentinel," *The Submarine Review*, Monterey, CA: US Navy, Naval Postgraduate School (April 2011): 78, http://www.nps.edu /Academic/Schools/Departments/USW/Documents/HOWARDAPR2011 .pdf.
18. William P. Gruner, "U.S. Pacific Submarines in World War II," Historic Naval Ships Association, 2010, www.hnsa.org/doc/subsinpacific.htm.
19. Gordon Corera, "Why Is Google in Love with Bletchley Park?," *BBC News*, November 16, 2011, www.bbc.co.uk/news/magazine-15739984.
20. "The Significance of the Impact of Bletchley Park on WWII, the Twentieth Century and the Way We All Live Today," Bletchley Park, National Codes Centre, http://www.bletchleypark.org.uk/content/hist/history/quotes.rhtm.
21. Ibid.
22. David Kahn, *The Codebreakers*, New York: Macmillan, 1967.
23. "Machines behind the Codes: Enigma," Bletchley Park, National Codes Centre, www.bletchleypark.org.uk/content/machines.rhtm.
24. Mavis Batey, *Dilly: The Man Who Broke Enigmas*, London: Biteback Publishing, 2011.
25. Jack Copeland, *Colossus: The First Electronic Computer*, Oxford, UK: Oxford University Press, 2006, www.colossus-computer.com/colossus1.html.
26. Ibid.
27. Ibid.

6 Computers and Computing

1. "Charles Babbage (Dec. 1791–Oct. 1871): Biography, Computer Models and Inventions," www.charlesbabbage.net.
2. "IBM Archives: Valuable Resources on IBM's History," www.ibm.com./ibm /history.
3. Ibid.; Mary Bellis, "Herman Hollerith: Punch Cards," http://inventors.about .com/library/inventors/blhollerith.htm.
4. Jane Smiley, *The Man Who Invented the Computer: The Biography of John Atanasoff*, New York: Doubleday, 2010.

5. "The First Computer at Penn: ENIAC," Penn, the University of Pennsylvania, www.upenn.edu/spotlights/first-computer-penn-eniac.

6. "Milestones: Whirlwind Computer," IEEE Global History Network, http://www.ieeeghn.org/wiki/index.php/Milestones:Whirlwind_Computer.

7. "National Archive for the History of Computing: Ferranti Ltd. Collection," ELGAR: Electronic Gateway to Archives at Rowlands, The John Rylands University Library, University of Manchester, UK, http://archives.li.man.ac.uk/ead/html/gb133nahc-fer-p1.shtml.

8. Simon Lavington, "Elliott Brothers (London) Ltd. and Elliott-Automation," Our Computer Heritage, http://www.ourcomputerheritage.org/Elliott%20company%20rev.pdf.

9. "About NCR: Company Overview, History / Timeline," www.ncr.com/about-ncr/company-overview/history-timeline.

10. "Father of British Computing, Sir Maurice Wilkes Dies," *BBC News*, Technology, November 30, 2010.

11. Peter J. Bird, *Leo: The First Business Computer*, Wokingham: Hasler Publishing, 1994.

12. Ibid.

13. Amanda Moore, "The Leo Computer and J. Lyons & Co.," November 11, 2011, www.intriguing-history.com/leo-computer-jlyons-co.

14. "Control Data Systems, Inc. History," www.fundinguniverse.com/company-histories/control-data-systems-inc-history/.

15. Vera B. Karpova, Leonid E. Karpov, *History of the Creation of BESM: The First Computer of S. A. Lebedev Institute of Precise Mechanics and Computer Engineering*, Heidelberg: Springer, 2006.

16. "Digital Equipment Corporation History," www.fundinguniverse.com/company-histories/digital-equipment-corporation-history/.

17. "The HP Computer Museum: Exhibit," www.hpmuseum.net/.

18. Ibid.

19. "Micral N of François Gernelle,"' www.history-computer.com/Modern Computer/Personal/Micral.html.

20. C. P. Thacker, E. M. McCreight, B. W. Lampson, R. F. Sproull, and D. R. Boggs, *Alto: A Personal Computer*, Palo Alto, CA: Palo Alto Research Center, Xerox Corporation, 1979.

21. William Thomas Sanderson, "The Virtual Altair Museum," www.virtualaltair.com.

22. "MITS Altair 8800," PC History, www.pc-history.org/altair.htm.

23. "IMSAI 8800," PC History, www.pc-history.org/imsai.htm.

24. Larry [portcommodore.com], "Chronological History of Commodore Computer," www.commodore.ca/history.

25. Ibid.

26. Ibid.

27. Andy F. Mesa, "Apple History Timeline," http://applemuseum.bott.org/sections/history. In June 2012, Sotheby's, the leading Auctioneers, sold a

handmade Apple I, with motherboard and a BASIC software manual, at a whopping $374,500 in an auction.

28. Agam Shah, "The Apple II from Commodore? It Almost Happened," Computerworld, December 12, 2007, www.computerworld.com/s/article /9052698.

29. "Osborne 1," www.history-computer.com/ModernComputer/Personal/Osborne .html.

30. Ibid.

31. The IBM Personal Computer, IBM Archives, www.ibm.com/ibm/history.

32. Ibid.

33. Ibid.

34. Ibid.

35. T. R. N. Rao and Subhash Kak, ed., Computing Science in Ancient India, Lafayette, LA: University of South Western Louisiana, 2000.

36. Mary Bellis, "Microsoft: History of a Computing Giant," http://inventors .about.com/od/CorporateProfiles/p/Microsoft-History.htm.

37. "The History of Microsoft," Pie Software Inc., July 9, 2001, www.piesof twareinc.co.uk/textonly/microsoft.html.

7 Media Recorders/Players, Mobile Phones, Smart Devices, and Tablets

1. Oberlin Smith, "Some Possible Forms of Phonograph," Electrical World, September 8, 1888.

2. "Valdemar Poulsen: Telegraphone," Lemelson MIT Program, August 2003, http://web.mit.edu/invent/iow/poulsen.html.

3. "Telegraphone," IEEE Global History Network, www.ieeeghn.org/wiki /index.php/telegraphone.

4. "History of Compact Cassette," Vintage Cassettes, 2012, www.vintage cassettes.com/_history/history.htm.

5. Nick Smith, "Classic Projects—Sony Walkman," Engineering & Technology Magazine 7, no. 10, October 22, 2012.

6. Robert R. Phillips, "First-Hand: Bing Crosby and the Recording Revolution," IEEE Global History Network, www.ieeeghn.org/wiki/index .php/First-Hand:Bing_Crosby_and_the_Recording_Revolution.

7. Ibid.

8. "Sony Corp. Info: Corporate History," Sony Corporation, 2013, www .sony.net/SonyInfo/CorporateInfo/History/history.html.

9. Fraunhofer Gesselshaft, "The Mp3 History—Fraunhofer Institute for Integrated Circuits," www.iis.fraunhofer.de/en/bf/amm/mp3history.html.

10. Ibid.

11. Daily Mail Reporter, "Is This the World's First Cell Phone? Film from 1938 Shows Woman Talking on a Wireless Device," Daily Mail, London, March 31, 2013.

12. "1946 First Mobile Telephone Call—AT&T," AT&T Labs, www.corp.att .com/attlabs/reputation/timeline/46mobile.html.
13. Ibid.
14. Ibid.
15. "Introduction to Research in Motion (RIM)," www.blackberry.com/select /get_the_facts/pdfs/rim/rim_history.pdf.
16. Ibid.
17. "Samsung History," Samsung Mobiles, www.samsung-mobiles.net/history-of -samsung.html.
18. Glen Sanford, "Apple History: iPhone," http://apple-history.com/iphone.
19. Ibid.
20. "Tablet," TechTerms.com, www.techterms.com/definition/tablet.
21. Laura June, "The Apple Tablet: A Complete History, Supposedly," Engadget, January 26, 2010, www.engadget.com/2010/01/26/the-apple-tablet-a-complete -history-supposedly.
22. Glen Sanford, "Apple History: iPad," http://apple-history.com/ipad.

8 Computer Networks and the Internet

1. "History," DARPA, www.darpa.mil/About/History/History.aspx.
2. "The Arpanet: Forerunner of Today's Internet," Raytheon BBN Technologies, 2013, www.bbn.com/about/timeline/arpanet.
3. Ibid.
4. Janelle Nanos, "Return to sender," *Boston Magazine*, June 2012, www .bostonmagazine.com/2012/05/shiva-ayyaduri-email-us-postal-service.
5. Dick Craddock, Group Program Manager, Windows Live Hotmail, "A Short History of Hotmail," Inside Windows Live (blog), January 6, 2010.
6. "The Birth of the Web," CERN, Geneva, Switzerland, 2013, http://home .web.cern.ch/about/birth-web.
7. Ibid.
8. "About NCSA Mosaic," University of Illinois, 2013, www.ncsa.illinois .edu/Projects/mosaic.html.
9. Eric Sink, "Memoirs from the Browser Wars," 2013, www.ericsink.com /Browser_Wars.html.
10. http://pressroom.yahoo.net/pr/ycorp/history.aspx.
11. "Our History in Depth," Google, www.google.com/about/company/history.
12. Ibid.
13. Sean Fanning, "Google Gets Ungoogleable off Sweden's New Word List," *BBC News*, March 26, 2013.
14. Claire Cann Miller, "Web Search at Crossroads, and Relevance Is the Quest Now," *New York Times*, April 4, 2013.
15. Vincent G. Cerf, "Why the Internet Works and How to Break It," *Economic Times*, New Delhi, January 28, 2013.
16. "Cyber—Security: The Digital Arms Trade," *The Economist*, March 30, 2013.

17. "Sun Microsystems, Inc. History," *International Directory of Company Histories*, Vol. 30, St. James Press, 2000, www.fundinguniverse.com/company-histories /sun-microsystems-inc-history.
18. "Cisco Systems, Inc. History," *International Directory of Company Histories*, Vol. 34, St. James Press, 2000, www.fundinguniverse.com/company-histories /cisco-systems-inc-history.
19. "Tomorrow Starts Here: What Happens When We Wake Up the World," Cisco, www.cisco.com/tomorrowstartshere.
20. "History of Innovation: Company Background," Juniper Networks, www .juniper.net/us/en/company/profile/history/.

9 Chips" and Displays

1. "Greenleaf W. Pickard: Biography," IEEE Global History Network, www .ieeeghn.org/wiki/index.php/Greenleaf_W._Pickard.
2. Bartholomew Lee, "How Dunwoody's Chunk of 'Coal' saved both de Forest and Marconi," *AWA Review* 22 (2009): 1.
3. Mark R. Pinto, William F. Brinkman, and William W. Troutman "The Transistor's Discovery and What's Ahead," Bell Laboratories, Lucent Technologies, www.imec.be/essderc/papers-97/322.pdf.
4. "Transistorized! The History of the Invention of the Transistor," ScienCentral Inc. and The American Institute of Physics, 1999, www.pbs .org/transistor/album1/index.html.
5. William Shockley, "The Path to the Conception of the Junction Transistor," *IEEE Transactions on Electronic Devices* ED-23, no. 7 (July 1976): 597–620.
6. William F. Brinkman, Douglas E. Haggan, and William W. Troutman, "A History of the Invention of the Transistor and Where It Will Lead Us," *IEEE Journal of Solid State Circuits* 32, no. 12 (1997): 1858–1864.
7. "The Chip That Jack Built—the Answer to a Problem," Texas Instruments Inc., www.ti.com/corp/docs/kilbyctr/jackbuilt.shtml.
8. Ibid.
9. "National Semiconductor Corporation History," *International Directory of Company Histories* Vol. 69, St. James Press, 2005, www.fundinguniverse .com/company-histories/national-semiconductor-corporation-history.
10. Ibid.
11. "Intel Timeline: A History of Innovation," www.intel.com/content/www/us /en/history/historic-timeline.html.
12. Ibid.
13. Ibid.
14. "Chipping In," *The Economist*, July 14, 2012.
15. Richard Stevenson, "The LED's Dark Secret," IEEE Spectrum, August 1, 2009, http://spectrum.ieee.org/semiconductors/optoelectronics/the-leds-dark -secret.

16. Ibid.
17. "Overview," Universal Display Corporation, http://www.udcoled.com/default
    .asp?contentID=576.
18. Ibid.
19. Benjamin Gross, "How RCA lost the LCD," IEEE Spectrum, November 1,
    2012, http://spectrum.ieee.org/consumer-electronics/audiovideo/how-rca-lost
    -the-lcd.
20. Ibid.

Appendix 1: Company Histories in Brief: The Pioneers of Telegraphy, Telephony, Wireless, Radio, and Television

1. Charles L. Brown, "Bell System History," Beatrice Companies Inc. www
   .beatriceco.com/bti/porticus/bell/bellsystem_history.html.
2. "Our History," British Telecom (BT), www.btplc.com/Thegroup/BTshistory/.
3. Centre for Business History and L. M. Ericsson, "The History of Ericsson,"
   www.ericssonhistory.com.
4. "Marconi's Wireless Telegraph Co.," Grace's Guide, 2007, www.gracesguide
   .co.uk/Marconi's_Wireless_Telegraph_Co.
5. Ron Ramirez, "The History of Philco," March 12, 2012, www.philcora-
   dio.com/history.
6. Koninklijke Philips N.V., "Philips History: Our Heritage," www.philips.com
   /about/company/history.
7. "Philips to Transfer Its Audio, Video, Multimedia and Accessories
   Business to Funai,' www.newscenter.philips.com/main/corpcomms/news
   /press/2013/20130129-philips-transfers-lifestyle-entertainment-business
   -to-funai.wpd#.UgDEkT66bIU.
8. Siemens AG, "Siemens History," www.siemens.com/history. R. Freshwater,
   "History of Siemens & Halske," 1998, www.britishtelephones.com/histsie
   .htm.
9. Telefunken AG, "History of Telefunken 1903–1922," www.telefunken
   .com/company/history. Oliver Archut, "History of Telefunken AG," www
   .tab-funkenwerk.com/id42.html.
10. Stephen B. Adams, Orville R. Butler, *Manufacturing the Future: A History of
    Western Electric*, New York: Cambridge University Press, 1999.
11. Maury Klein, *The Life and Legend of Jay Gould*, Baltimore, MD: Johns Hopkins
    University Press, 1997.
12. "Our Rich History," Western Union, Western Union Holdings Inc., 2012,
    http://corporate.westernunion.com/History.html.
13. Thomas H. White, "United States Early Radio History: Big Business & Radio
    (1915–1922)," http://earlyradiohistory.us/sec017.htm.
14. "Westinghouse Electric Corporation History," *International Directory of
    Company Histories*, Vol. 12, St. James Press, 1996, www.fundinguniverse.com
    /company-histories/westinghouse-electric-corporation-history.

Appendix 2: Principal Entities Associated with Thomas Alva Edison and Their Timeline

* "Edison Companies," http://edison.rutgers.edu/list.htm.

Appendix 4: Principal Entities Associated with Guglielmo Marconi and Their Timeline

* Entities in italics were not started by Marconi and/or his companies but became associated later.

Appendix 5: Timeline—Offshoring of Semiconductor Assembly

* This timeline is only a representative list and may not be construed as a complete and comprehensive listing.

Index

Printed in the United States
By Bookmasters